REA: THE LEADER IN CLEP® TEST PREP

CLEP® COLLEGE MATHEMATICS

Mel Friedman, M.S.

and

Stu Schwartz

Research & Education Association

Visit our website at: www.rea.com

Research & Education Association
61 Ethel Road West
Piscataway, New Jersey 08854
E-mail: info@rea.com

CLEP® College Mathematics with Online Practice Exams

Printed in the United States of America

Library of Congress Control Number 2015956926

ISBN-13: 978-0-7386-1205-8
ISBN-10: 0-7386-1205-7

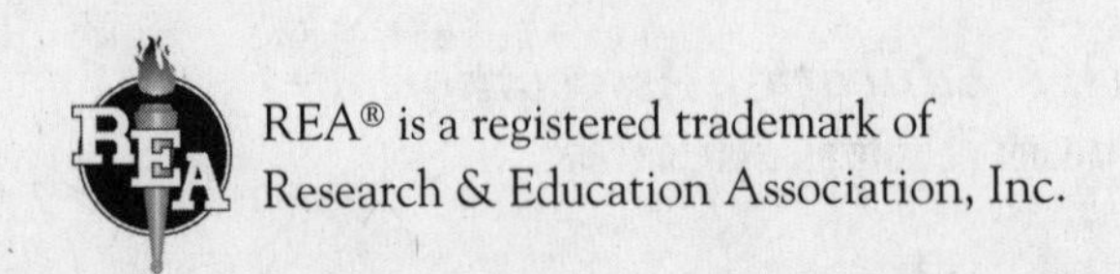

K16

CONTENTS

CHAPTER 3

CHAPTER 4

CHAPTER 8

ABOUT OUR AUTHORS

Mel Friedman, M.S., has a diverse background in mathematics, having taught both high school and college-level mathematics courses. Mr. Friedman was awarded his B.A. in Mathematics from Rutgers University and received his M.S. in Mathematics (with honors) from Fairleigh Dickinson University. A former adjunct professor at Kutztown University of Pennsylvania, Mr. Friedman also developed test items for Educational Testing Service and ACT, Inc.

Stu Schwartz has been teaching mathematics since 1973. For 35 years he taught in the Wissahickon School District, in Ambler, Pennsylvania, specializing in AP Calculus AB and BC and AP Statistics. Mr. Schwartz received his B.S. degree in Mathematics from Temple University, Philadelphia. Mr. Schwartz was a 2002 recipient of the Presidential Award for Excellence in Mathematics Teaching and also won the 2007 Outstanding Educator of the Year Award for the Wissahickon School District.

Mr. Schwartz's website, *www.mastermathmentor.com*, is geared toward helping educators teach AP Calculus, AP Statistics, and other math courses. Mr. Schwartz is always looking for ways to provide teachers with new and innovative teaching materials, believing that it should be the goal of every math teacher not only to teach students mathematics, but also to find joy and beauty in math as well.

ABOUT REA

Founded in 1959, Research & Education Association (REA) is dedicated to publishing the finest and most effective educational materials—including study guides and test preps—for students of all ages. Today, REA's wide-ranging catalog is a leading resource for students, teachers, and other professionals. Visit *www.rea.com* to see a complete listing of all our titles.

ACKNOWLEDGMENTS

We would like to thank Pam Weston, Publisher, for setting the quality standards for production integrity and managing the publication to completion; John Paul Cording, Vice President, Technology, for coordinating the design and development of the REA Study Center; Larry B. Kling, Vice President, Editorial, for overall direction; Diane Goldschmidt, Managing Editor, for coordinating development of this edition; Transcend Creative Services for typesetting this edition; and Eve Grinnell, Graphic Designer, for designing our cover.

In addition, we would like to thank Mary Berlinghieri Willi for technically reviewing the manuscript, Linda Robbian for copyediting, and Ellen Gong for proofreading.

CHAPTER 1

Passing the CLEP College Mathematics Exam

CHAPTER 1

PASSING THE CLEP COLLEGE MATHEMATICS EXAM

Congratulations! You're joining the millions of people who have discovered the value and educational advantage offered by the College Board's College-Level Examination Program, or CLEP. This test prep focuses on what you need to know to succeed on the CLEP College Mathematics exam, and will help you earn the college credit you deserve while reducing your tuition costs.

GETTING STARTED

There are many different ways to prepare for a CLEP exam. What's best for you depends on how much time you have to study and how comfortable you are with the subject matter. To score your highest, you need a system that can be customized to fit you: your schedule, your learning style, and your current level of knowledge.

This book, and the online tools that come with it, allow you to create a personalized study plan through three simple steps: assessment of your knowledge, targeted review of exam content, and reinforcement in the areas where you need the most help.

Let's get started and see how this system works.

Test Yourself & Get Feedback	Assess your strengths and weaknesses. The score report from your online diagnostic exam gives you a fast way to pinpoint what you already know and where you need to spend more time studying.
Review with the Book	Armed with your diagnostic score report, review the parts of the book where you're weak and study the answer explanations for the test questions you answered incorrectly.
Ensure You're Ready for Test Day	After you've finished reviewing with the book, take our full-length practice tests. Review your score reports and re-study any topics you missed. We give you two full-length practice tests to ensure you're confident and ready for test day.

THE REA STUDY CENTER

The best way to personalize your study plan is to get feedback on what you know and what you don't know. At the online REA Study Center, you can access two types of assessment: a diagnostic exam and full-length practice exams. Each of these tools provides true-to-format questions and delivers a detailed score report that follows the topics set by the College Board.

Diagnostic Exam

Before you begin your review with the book, take the online diagnostic exam. Use your score report to help evaluate your overall understanding of the subject, so you can focus your study on the topics where you need the most review.

Full-Length Practice Exams

Our full-length practice tests give you the most complete picture of your strengths and weaknesses. After you've finished reviewing with the book, test what you've learned by taking the first of the two online practice exams. Review your score report, then go back and study any topics you missed. Take the second practice test to ensure you have mastered the material and are ready for test day.

If you're studying and don't have Internet access, you can take the printed tests in the book. These are the same practice tests offered at the REA Study Center, but without the added benefits of timed testing conditions and diagnostic score reports. Because the actual exam is Internet-based, we recommend you take at least one practice test online to simulate test-day conditions.

AN OVERVIEW OF THE EXAM

The CLEP College Mathematics exam consists of approximately 60 questions to be answered in 90 minutes.

The exam covers the material one would find in an introductory college-level class for nonmathematics majors in fields not requiring knowledge of advanced mathematics.

The approximate breakdown of topics is as follows:

20% Algebra and Functions

10% Counting and Probability

15% Data Analysis and Statistics

20% Financial Mathematics

10% Geometry

15% Logic and Sets

10% Numbers

Scientific Calculator

A scientific (nongraphing) calculator is integrated into the exam software, and is available to students during the entire testing time. Students are expected to know how and when to make appropriate use of the calculator. The scientific calculator, together with a brief video tutorial, is available to students as a free download for a 30-day trial period. Visit *www.collegeboard.org/clep* for more information.

ALL ABOUT THE CLEP PROGRAM

What is the CLEP?

CLEP is the most widely accepted credit-by-examination program in North America. The CLEP program's 33 exams span five subject areas. The exams assess the material commonly required in an introductory-level college course. Examinees can earn from three to twelve credits at more than 2,900 colleges and universities in the U.S. and Canada. For a complete list of the CLEP subject examinations offered, visit the College Board website: *www.collegeboard.org/clep*.

Who takes CLEP exams?

CLEP exams are typically taken by people who have acquired knowledge outside the classroom and who wish to bypass certain college courses and earn college credit. The CLEP program is designed to reward examinees for learning—no matter where or how that knowledge was acquired.

Although most CLEP examinees are adults returning to college, many graduating high school seniors, enrolled college students, military personnel, veterans, and international students take CLEP exams to earn college credit or to demonstrate their ability to perform at the college level. There are no prerequisites, such as age or educational status, for taking CLEP examinations. However, because policies on granting credits vary among colleges, you should contact the particular institution from which you wish to receive CLEP credit.

How is my CLEP score determined?

Your CLEP score is based on two calculations. First, your CLEP raw score is figured; this is just the total number of test items you answer correctly. After the test is administered, your raw score is converted to a scaled score through a process called *equating*. Equating adjusts for minor variations in difficulty across test forms and among test items, and ensures that your score accurately represents your performance on the exam regardless of when or where you take it, or on how well others perform on the same test form.

Your scaled score is the number your college will use to determine if you've performed well enough to earn college credit. Scaled scores for the CLEP exams are delivered on a 20-80 scale. Institutions can set their own scores for granting college credit, but a good passing estimate (based on recommendations from the American Council on Education) is generally a scaled score of 50, which usually requires getting roughly 66% of the questions correct.

For more information on scoring, contact the institution where you wish to be awarded the credit.

Who administers the exam?

CLEP exams are developed by the College Board, administered by Educational Testing Service (ETS), and involve the assistance of educators from throughout the United States. The test development process is designed and implemented to ensure that the content and difficulty level of the test are appropriate.

When and where is the exam given?

CLEP exams are administered year-round at more than 1,200 test centers in the United States and can be arranged for candidates abroad on request. To find the test center nearest you and to register for the exam, contact the CLEP Program:

CLEP Services
P.O. Box 6600
Princeton, NJ 08541-6600
Phone: (800) 257-9558 (8 A.M. to 6 P.M. ET)
Fax: (610) 628-3726
Website: *www.collegeboard.org/clep*

The CLEP iBT Platform

To improve the testing experience for both institutions and test-takers, the College Board's CLEP Program has transitioned its 33 exams from the eCBT platform to an Internet-based testing (iBT) platform. All CLEP test-takers may now register for exams and manage their personal account information through the "My Account" feature on the CLEP website. This new feature simplifies the registration process and automatically downloads all pertinent information about the test session, making for a more streamlined check-in.

OPTIONS FOR MILITARY PERSONNEL AND VETERANS

CLEP exams are available free of charge to eligible military personnel and eligible civilian employees. All the CLEP exams are available at test centers on college campuses and military bases. Contact your Educational Services Officer or Navy College Education Specialist for more information. Visit the DANTES or College Board websites for details about CLEP opportunities for military personnel.

Eligible U.S. veterans can claim reimbursement for CLEP exams and administration fees pursuant to provisions of the Veterans Benefits Improvement Act of 2004. For details on eligibility and submitting a claim for reimbursement, visit the U.S. Department of Veterans Affairs website at *www.gibill.va.gov*.

CLEP can be used in conjunction with the Post-9/11 GI Bill, which applies to veterans returning from the Iraq and Afghanistan theaters of operation. Because the GI Bill provides tuition for up to 36 months, earning college credits with CLEP exams expedites academic progress and degree completion within the funded timeframe.

SSD ACCOMMODATIONS FOR CANDIDATES WITH DISABILITIES

Many test candidates qualify for extra time to take the CLEP exams, but you must make these arrangements in advance. For information, contact:

College Board Services for Students with Disabilities
P.O. Box 8060
Mt. Vernon, Illinois 62864-0060

Phone: (609) 771-7137 (Monday through Friday, 8 A.M. to 6 P.M. ET)
TTY: (609) 882-4118
Fax: (866) 360-0114
Website: *http://student.collegeboard.org/services-for-students-with-disabilities*
E-mail: ssd@info.collegeboard.org

6-WEEK STUDY PLAN

Although our study plan is designed to be used in the six weeks before your exam, it can be condensed to three weeks by combining each two-week period into one.

Be sure to set aside enough time—at least two hours each day—to study. The more time you spend studying, the more prepared and relaxed you will feel on the day of the exam.

Week	Activity
1	Take the Diagnostic Exam at the online REA Study Center. Your score report will identify topics where you need the most review.
2–4	Study the review, focusing on the topics you missed (or were unsure of) on the Diagnostic Exam.
5	Take Practice Test 1 at the REA Study Center. Review your score report and re-study any topics you missed.
6	Take Practice Test 2 at the REA Study Center to see how much your score has improved. If you still got a few questions wrong, go back to the review and study any topics you may have missed.

TEST-TAKING TIPS

Know the format of the test. Familiarize yourself with the CLEP computer screen beforehand by logging on to the College Board website. Waiting until test day to see what it looks like in the pretest tutorial risks injecting needless anxiety into your testing experience. Also, familiarizing yourself with the directions and format of the exam will save you valuable time on the day of the actual test.

Read all the questions—completely. Make sure you understand each question before looking for the right answer. Reread the question if it doesn't make sense.

Read all of the answers to a question. Just because you think you found the correct response right away, do not assume that it's the best answer. The last answer choice might be the correct answer.

Work quickly and steadily. You will have 90 minutes to answer 60 questions, so work quickly and steadily. Taking the timed practice tests online will help you learn how to budget your time.

Use the process of elimination. Stumped by a question? Don't make a random guess. Eliminate as many of the answer choices as possible. By eliminating just one answer choice, you give yourself a better chance of getting the item correct, since there will only be three choices left from which to make your guess. Remember, your score is based only on the number of questions you answer correctly.

Don't waste time! Don't spend too much time on any one question. Your time is limited so pacing yourself is very important. Work on the easier questions first. Skip the difficult questions and go back to them if you have the time.

Look for clues to answers in other questions. If you skip a question you don't know the answer to, you might find a clue to the answer elsewhere on the test.

Be sure that your answer registers before you go to the next item. Look at the screen to see that your mouse-click causes the pointer to darken the proper oval. If your answer doesn't register, you won't get credit for that question.

THE DAY OF THE EXAM

On test day, you should wake up early (after a good night's rest, of course) and have breakfast. Dress comfortably so you are not distracted by being too hot or too cold while taking the test. (Note that "hoodies" are not allowed.) Arrive at the test center early. This will allow you to collect your thoughts and relax before the test, and it will also spare you the anxiety that comes with being late.

Before you leave for the test center, make sure you have your admission form and another form of identification, which must contain a recent photograph, your name, and signature (i.e., driver's license, student identification card, or current alien registration card). You may not wear a digital watch (wrist or pocket), alarm watch, or wristwatch camera. In addition, no cell phones, dictionaries, textbooks, notebooks, briefcases, or packages will be permitted, and drinking, smoking, and eating are prohibited.

Good luck on the CLEP College Mathematics exam!

CHAPTER 2

Numbers

CHAPTER 2

NUMBERS

Real numbers provide the basis for most precalculus mathematics topics.

> **Real numbers** are all of the numbers on the **number line** (see Figure 2-1).

In fact, a nice way to visualize real numbers is that they can be put in a one-to-one correspondence with the set of all points on a line. Real numbers include positives, negatives, square roots, π (pi), and just about any number you have ever encountered.

Figure 2-1

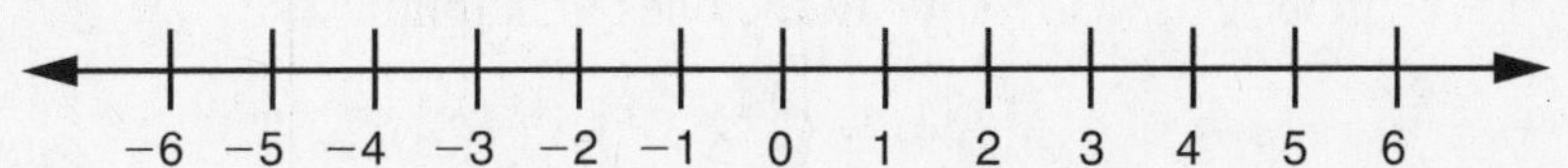

PROPERTIES OF REAL NUMBERS

Similar to the Laws of Set Operations, real numbers have several properties that you should know. You have used some of these properties ever since you could count. You intuitively knew that $3 + 2$ gives the same result as $2 + 3$, or that if you add 0 to a number it remains unchanged. These properties deal with addition and multiplication. They do not work for subtraction and division. For example, you also intuitively know that $3 - 2$ is not the same as $2 - 3$.

Perhaps you are not familiar with the names of these properties. The following list provides the names for these properties. Learn them—you will encounter these property names in the next chapter and on mathematics tests. The examples use the numbers 2, 3, and 4, but the rules apply to any real numbers.

COMMUTATIVE PROPERTY

The numbers *commute*, or move:

Addition $2 + 3 = 3 + 2$

Multiplication $2 \times 3 = 3 \times 2$

ASSOCIATIVE PROPERTY

The numbers can be grouped, or *associated*, in any order:

Addition $2 + (3 + 4) = (2 + 3) + 4$

Multiplication $2 \times (3 \times 4) = (2 \times 3) \times 4$

Later in this chapter, you will see that operations in parentheses are always done first, but the associative property says that you can move the parentheses and it won't make a difference.

DISTRIBUTIVE PROPERTY

The first number gets *distributed* to the ones in parentheses:

$$2 \times (3 + 4) = (2 \times 3) + (2 \times 4)$$

The following properties have to do with the special numbers 0 and 1:

IDENTITY PROPERTY

Adding 0 or multiplying by 1 doesn't change the original value:

Addition $3 + 0 = 3$

Multiplication $3 \times 1 = 3$

INVERSE PROPERTY

The *inverse* of addition is subtraction and the *inverse* of multiplication is division:

Additive Inverse $3 + (-3) = 0$

Multiplicative Inverse $3 \times \frac{1}{3} = 1$

Note that the multiplicative inverse doesn't work for 0 because division by 0 is not defined.

COMPONENTS OF REAL NUMBERS

The set of all real numbers (designated as R) has various components:

$N = \{1, 2, 3, \ldots\}$, the set of all **natural numbers**

$W = \{0, 1, 2, 3, \ldots\}$, the set of all **whole numbers**

$I = \{\ldots, -3, -2, -1, 0, 1, 2, 3, \ldots\}$, the set of all **integers**

$Q = \left\{\frac{a}{b} | \, a, b \in I \text{ and } b \neq 0\}\right\}$, the set of all **rational numbers**

$S = \{x \mid x$ has a decimal that is nonterminating and does not have a repeating block$\}$, the set of all **irrational numbers**.

It is obvious that $N \subseteq W$, $W \subseteq I$, and $I \subseteq Q$, but a similar relationship does not hold between Q and S. More specifically, the decimal names for elements of Q are either (1) terminating or (2) nonterminating with a repeating block.

Examples of rational numbers include $\frac{1}{2} = .5$ and $\frac{1}{3} = .333\ldots$

This means that Q and S have no common elements.

Examples of irrational numbers include $.101001000\ldots$, π, and $\sqrt{2}$.

All real numbers are normally represented by R and $R = Q \cup S$. This means that every real number is either rational or irrational.

FRACTIONS

All rational numbers can be displayed as **fractions**, which consist of a numerator (on the top) and a denominator (on the bottom).

Proper fractions are numbers between -1 and $+1$; the numerator is less than the denominator. Examples of proper fractions are $\frac{1}{2}$, $\frac{3}{4}$, and $\frac{17}{19}$.

Improper fractions are all other rational numbers; the numerator is greater than or equal to the denominator.

Improper fractions are also called mixed numbers because they can be written as a whole number with a fractional part. Examples of improper fractions are $\frac{2}{1}$, $\frac{4}{3}$, and $\frac{19}{17}$. The first of these is actually a whole number (2); the others are equivalent to the mixed numbers $1\frac{1}{3}$ and $1\frac{2}{17}$, respectively.

ODD AND EVEN NUMBERS

When dealing with odd and even numbers, keep in mind the following:

Adding:

even + even = even

odd + odd = even

even + odd = odd

Multiplying:

even × even = even

even × odd = even

odd × odd = odd

FACTORS AND DIVISIBILITY NUMBERS

> Any counting number that divides into another number with no remainder is called a **factor** of that number.

The factors of 20 are 1, 2, 4, 5, 10, and 20.

> Any number that can be divided by another number with no remainder is called a **multiple** of that number.

Examples of multiples of 20 are 20, 40, 60, 80, etc.

PROBLEM

List the factors and multiples of 28.

SOLUTION

The factors of 28 are 1, 2, 4, 7, 14, and 28. Some multiples of 28 are 28, 56, 84, and 112. Note that the list of multiples is endless.

ABSOLUTE VALUE

> The **absolute value** of a number is represented by two vertical lines around the number and is equal to the positive value, regardless of sign.

The absolute value of a real number A is defined as follows:

$$|A| = \begin{cases} A \text{ if } A \geq 0 \\ -A \text{ if } A < 0 \end{cases}$$

Examples:

$$|5| = 5$$
$$|-8| = -(-8) = 8$$

Absolute values follow the given rules:

1. $|-A| = |A|$
2. $|A| \geq 0$, equality holding only if $A = 0$
3. $\left|\frac{A}{B}\right| = \frac{|A|}{|B|}, B \neq 0$
4. $|AB| = |A| \times |B|$
5. $|A|^2 = A^2$

PROBLEM

Calculate the value of each of the following expressions:

1. $||2 - 5| + 6 - 14|$
2. $|-5| \times |4| + \dfrac{|-12|}{4}$

SOLUTION

Before solving this problem, one must remember the order of operations: parentheses, multiplication and division, addition and subtraction.

1. $||-3| + 6 - 14| = |3 + 6 - 14| = |9 - 14| = |-5| = 5$
2. $(5 \times 4) + \dfrac{12}{4} = 20 + 3 = 23$

INTEGERS

There are various subsets of I, the set of all integers:

NEGATIVE INTEGERS

The set of integers starting with -1 and decreasing:

$$\{-1, -2, -3, \ldots\}.$$

EVEN INTEGERS

The set of integers divisible by 2:

$$\{\ldots, -4, -2, 0, 2, 4, 6, \ldots\}.$$

ODD INTEGERS

The set of integers not divisible by 2:

$$\{\ldots, -3, -1, 1, 3, 5, 7, \ldots\}.$$

CONSECUTIVE INTEGERS

The set of integers that differ by 1:

$$\{n, n + 1, n + 2, \ldots\} \ (n = \text{an integer}).$$

PRIME NUMBERS

The set of positive integers greater than 1 that are divisible only by 1 and themselves:

$$\{2, 3, 5, 7, 11, \ldots\}.$$

COMPOSITE NUMBERS

The set of integers, other than 0 and ± 1, that are not prime.

PROBLEMS

Classify each of the following numbers into as many different sets as possible.

Example: real, integer …

1. 0
2. 9
3. $\sqrt{6}$
4. $\frac{1}{2}$
5. $\frac{2}{3}$
6. 1.5
7. 11

SOLUTIONS

1. 0 is a real number, an integer, a whole number, and a rational number.
2. 9 is a real number, an odd number, a natural number, and a rational number.
3. $\sqrt{6}$ is a real number, and an irrational number.
4. $\frac{1}{2}$ is a real number, and a rational number.
5. $\frac{2}{3}$ is a real number, and a rational number.
6. 1.5 is a real number, a decimal, and a rational number.
7. 11 is a prime number, an odd number, a real number, a natural number, and a rational number.

INEQUALITIES

If x and y are real numbers, then one and only one of the following statements is true:

$$x > y, x = y, \text{ or } x < y.$$

This is the **order property of real numbers**.

If a, b, and c are real numbers, the following statements are true:

If $a < b$ and $b < c$, then $a < c$.

If $a > b$ and $b > c$, then $a > c$.

This is the **transitive property of inequalities**.

If a, b, and c are real numbers and $a > b$, then $a + c > b + c$ and $a - c > b - c$. This is the **addition property of inequality**.

> An **inequality** is a statement in which the value of one quantity or expression is greater than (>), less than (<), greater than or equal to (≥), less than or equal to (≤), or not equal to (≠) that of another.

Example:

5 > 4. This expression means that the value of 5 is greater than the value of 4.

The **graph of an inequality** in one variable is represented by either a ray or a line segment on the real number line.

The endpoint is not a solution if the variable is strictly less than or greater than a particular value. In those cases, the endpoint is indicated by an open circle.

Example:

$x > 2$

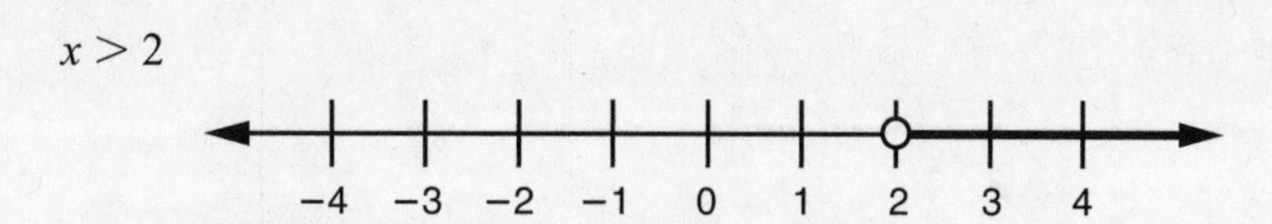

2 is not a solution and should be represented as shown.

The endpoint is a solution if the variable is either (1) less than or equal to or (2) greater than or equal to a particular value. In those cases, the endpoint is indicated by a closed circle.

Example:

$5 > x \geq 2$

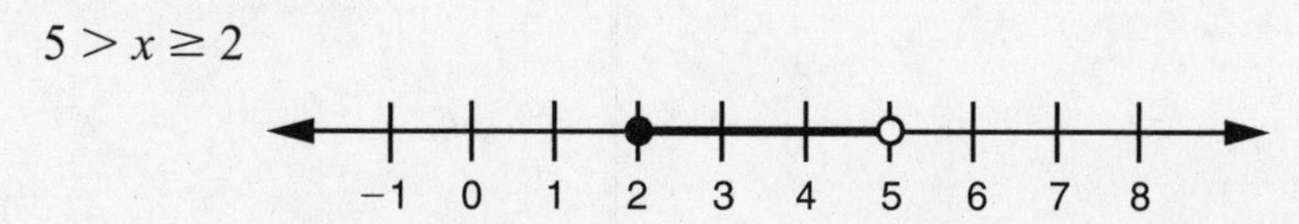

In this case, 2 is a solution and 5 is not a solution, and the solution should be represented as shown.

Example:

$x < 2$ or $x > 5$

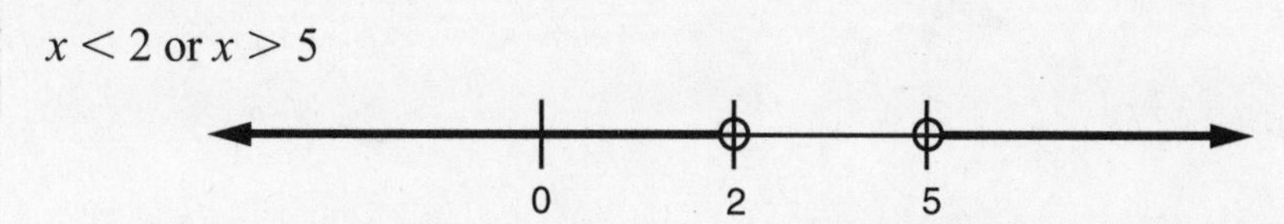

In this case, neither 2 nor 5 is a solution. Thus, an open circle must be shown at $x = 2$ and at $x = 5$.

Example:

$x \leq 2$ and $x \geq 5$

In this case, there is no solution. It is impossible for a number to be both no greater than 2 and no less than 5.

Example:

$x \geq 2$ or $x \leq 5$

In this case, the solution is all real numbers. Any number *must* belong to at least one of these inequalities. Some numbers, such as 3, belong to both inequalities. If you graph these inequalities separately, you will notice two rays going in opposite directions and which overlap between 2 and 5, inclusive.

Intervals on the number line represent sets of points that satisfy the conditions of an inequality.

> An **open** interval does not include any endpoints.

Example:

$\{x \mid x > -3\}$, read as "the set of values x such that $x > -3$." The graph would appear as:

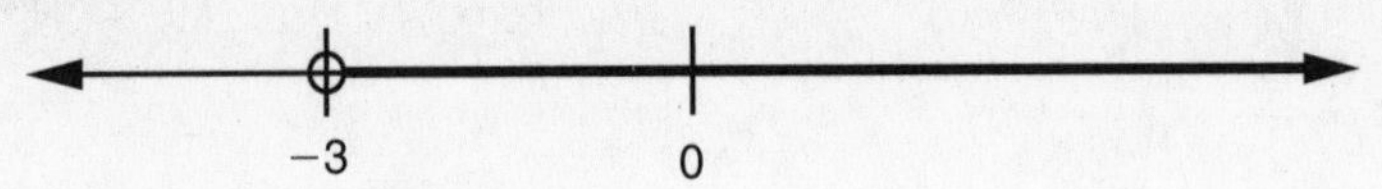

Example:

$\{x \mid x < 4\}$. The graph would appear as:

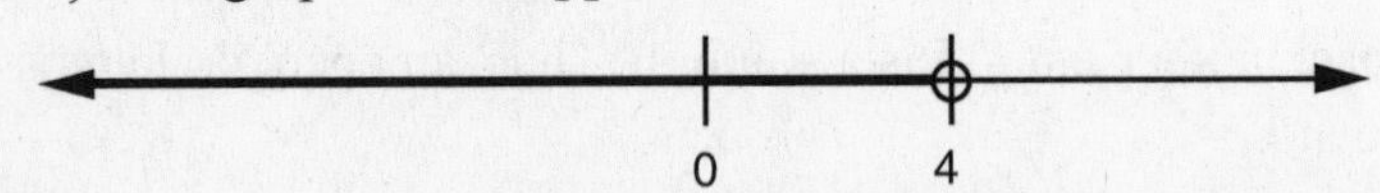

Example:

$\{x \mid 2 < x < 6\}$. The graph would appear as:

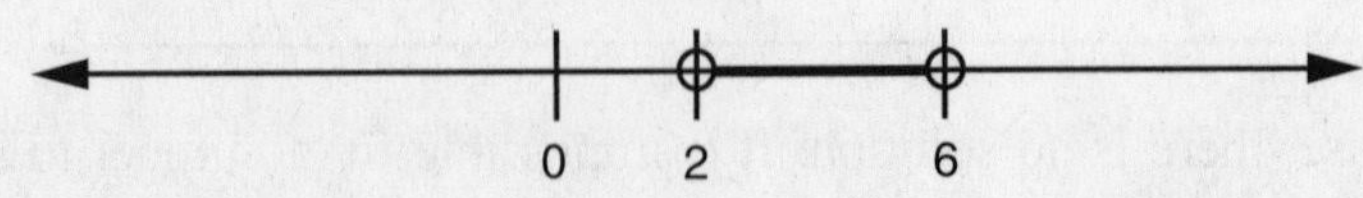

Example:

$\{x \mid x \text{ is any real number}\}$. The graph would appear as:

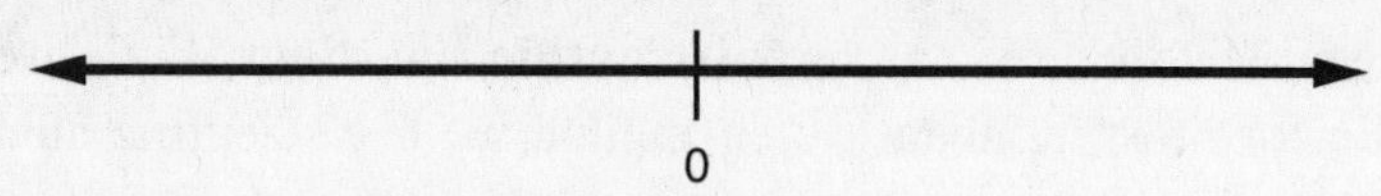

A **closed** interval includes two endpoints.

Example:

$\{x \mid -5 \leq x \leq 2\}$. The graph would appear as:

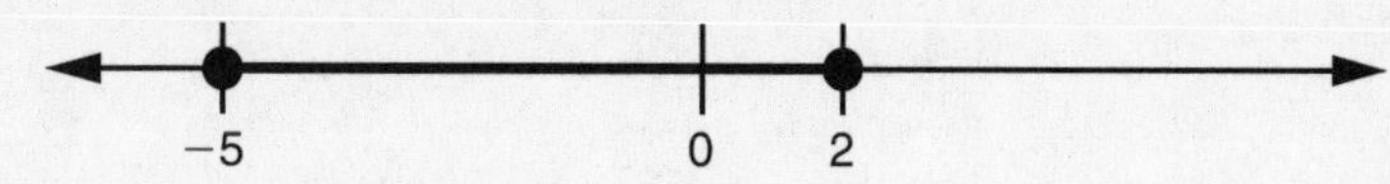

A **half-open** interval includes one endpoint.

Example:

$\{x \mid x \geq 3\}$. The graph would appear as:

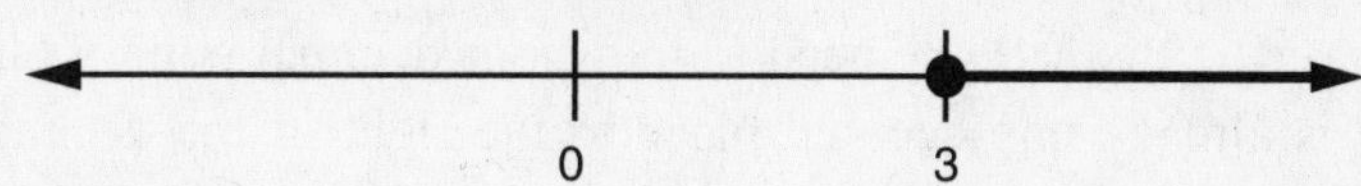

Example:

$\{x \mid x \leq 6\}$. The graph would appear as:

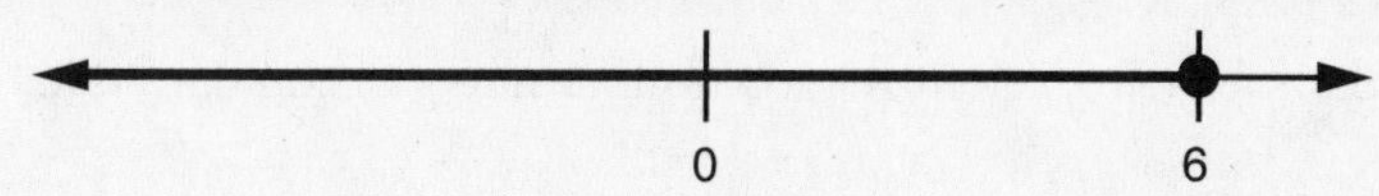

Example:

$\{x \mid -4 < x \leq -1\}$.

−4 −1 0

Example:

$\{x \mid -2 \leq x < 1\}$.

−2 0 1

SCIENTIFIC NOTATION

It is possible that some of the numbers or answers you encounter on the CLEP College Math exam may be in **scientific notation**. While we use standard notation for most numbers we encounter, we use scientific notation when working with numbers that are either very large or very small. The number of cells in the human body is estimated to be 150,000,000,000,000 (read 150 trillion) and is written as 1.5×10^{14}. The length of time it takes light to travel 1 meter is approximately 0.000000003 second and is written as 3×10^{-9}.

The form for a number in scientific notation is $a \times 10^n$ where $1 \leq a < 10$ and n is an integer.

SCIENTIFIC NOTATION TO STANDARD NOTATION

To convert a number in scientific notation to standard notation, we add zeros to it. If n is positive, we add zeros to the right. If n is negative, we add zeros to the left.

So to convert 1.2×10^5, we need to move our decimal point 5 places to the right. There is already one decimal place to the right of the decimal point so we add 4 zeros. So $1.2 \times 10^5 = 120000$. We usually write 120000 as 120,000, making it easier to read.

PROBLEMS

Change the following scientific notation numbers to standard notation.

(A) 3.7×10^4

(B) 9.92×10^2

(C) 1.005×10^9

SOLUTIONS

(A) $3.7 \times 10^4 = 370,000$

(B) $9.92 \times 10^2 = 992$

(C) $1.005 \times 10^9 = 1,005,000,000$

To convert 4.7×10^{-4}, we need to move our decimal point 4 places to the left. There is already 1 decimal place to the left of the decimal point so we add 3 zeros. So $4.7 \times 10^{-4} = 0.00047$.

PROBLEMS

Change the following scientific notation numbers to standard notation.

(A) 5.1×10^{-3}

(B) 2.27×10^{-6}

(C) 8.004×10^{-5}

(D) 4.2×10^{0}

SOLUTIONS

(A) $5.1 \times 10^{-3} = 0.0051$

(B) $2.27 \times 10^{-6} = 0.00000227$

(C) $8.004 \times 10^{-5} = 0.00008004$

(D) $4.2 \times 10^{0} = 4.2$

STANDARD NOTATION TO SCIENTIFIC NOTATION

To convert numbers to scientific notation, we focus on large numbers (numbers greater than 1) or small numbers (numbers less than 1).

Large numbers: To convert 34,320 to scientific notation, we place a caret after the first non-zero digit, the 3. We then count the number of decimal places from the caret to the real decimal point, which is at the end of the number. There are 4 decimal places. So our answer is $34{,}320 = 3.432 \times 10^{4}$.

If the number is 25,460,000,000, we place we place a caret after the first non-zero digit, the 2. We then count the number of decimal places from the caret to the real decimal point, which is at the end of the number. There are 10 decimal places. So our answer is $25{,}460{,}000{,}000 = 2.546 \times 10^{10}$.

Small numbers: To convert 0.000073 to scientific notation, place a caret after the first non-zero digit, the 7. We then count the number of decimal places from the caret to the real decimal point, which is to the left. There are 5 decimal places, so our answer is $0.000073 = 7.3 \times 10^{-5}$.

If the number is 0.01503, we place a caret after the first non-zero digit, the 1. We then count the number of decimal places from the caret to the real decimal point, which is to the left. There are 2 decimal places, so our answer is $0.01503 = 1.503 \times 10^{-2}$.

PROBLEMS

Change the following standard notation numbers to scientific notation.

(A) 600,000,000,000

(B) 2,450

(C) −10,080,000

(D) 0.79

(E) − 0.000005505

SOLUTIONS

(A) $600{,}000{,}000{,}000 = 6 \times 10^{11}$

(B) $2{,}450 = 2.45 \times 10^{3}$

(C) $-10{,}080{,}000 = -1.008 \times 10^{7}$

(D) $0.79 = 7.9 \times 10^{-1}$

(E) $-0.000005505 = -5.505 \times 10^{-6}$

MULTIPLICATION AND DIVISION OF NUMBERS IN SCIENTIFIC NOTATION

To multiply two numbers in scientific notation, we multiply the numbers and add the exponents of the 10's. So $(a \times 10^{m})(b \times 10^{n}) = ab \times 10^{m+n}$. This method works whether the exponents m and n are positive or negative. So $(4 \times 10^{4})(2 \times 10^{5}) = 4(2) \times 10^{4+5} = 8 \times 10^{9}$. If when we multiply a times b, it is possible that this number is greater than 10. If so, change it to scientific notation as well.

PROBLEMS

Multiply the following.

(A) $(3 \times 10^8)(1.4 \times 10^7)$

(B) $(-7 \times 10^4)(1.1 \times 10^{-6})$

(C) $(-1.2 \times 10^{-2})^2$

(D) $(-3 \times 10^{10})(-5 \times 10^{-4})$

SOLUTIONS

(A) $(3 \times 10^8)(1.4 \times 10^7) = 4.2 \times 10^{15}$

(B) $(-7 \times 10^4)(1.1 \times 10^{-6}) = -7.7 \times 10^{-2}$

(C) $(-1.2 \times 10^{-2})^2 = (-1.2 \times 10^{-2})(-1.2 \times 10^{-2}) = 1.44 \times 10^{-4}$

(D) $(-3 \times 10^{10})(-5 \times 10^{-4}) = 15 \times 10^6 = 1.5 \times 10^1 \times 10^6 = 1.5 \times 10^7$

To divide two numbers in scientific notation, we divide the numbers and subtract the exponents of the 10's (numerator exponent minus denominator exponent). So $\frac{a \times 10^m}{b \times 10^n} = \frac{a}{b} \times 10^{m-n}$. This method works whether the exponents m and n are positive or negative. So, $\frac{6 \times 10^9}{2 \times 10^4} = \frac{6}{2} \times 10^{9-4} = 3 \times 10^5$. When we divide a by b, it is possible that this number is less than 1. If so, change it to scientific notation as well.

PROBLEMS

Find the following.

(A) $\frac{3.2 \times 10^{15}}{2 \times 10^7}$

(B) $\frac{7.5 \times 10^4}{2.5 \times 10^{-2}}$

(C) $\frac{1.21 \times 10^{-3}}{-1.1 \times 10^{-6}}$

(D) $\frac{-2 \times 10^{-3}}{-8 \times 10^{-2}}$

SOLUTIONS

(A) $\dfrac{3.2\times10^{15}}{2\times10^{7}}=1.6\times10^{15-7}=1.6\times10^{8}$

(B) $\dfrac{7.5\times10^{4}}{2.5\times10^{-2}}=3\times10^{4-(-2)}=3\times10^{6}$

(C) $\dfrac{1.21\times10^{-3}}{-1.1\times10^{-6}}=-1.1\times10^{-3-(-6)}=-1.1\times10^{3}$

(D) $\dfrac{-2\times10^{-3}}{-8\times10^{-2}}=0.25\times10^{-3-(-2)}=2.5\times10^{-1}\times10^{-1}=2.5\times10^{-2}$

UNIT CONVERSIONS

Length and liquid measurement units are usually given in the Imperial (or Standard) system (which you are used to) or the Metric system (which is used in just about all countries except the United States). The relationships below are ones that you should know. If it is necessary to change from an imperial measurement to a metric measurement, you will be given the relationship (ex: 1 meter = 39.37 inches). There are basic relationships that you should know when you change from one unit to another.

Length

Imperial system

1 foot = 12 inches

1 yard = 3 feet

1 mile = 5,280 feet

Metric system

1 meter = 100 centimeters

1 meter = 1,000 millimeters

1 kilometer = 1,000 meters

Weight

Imperial system

1 pound = 16 ounces

1 ton = 2,000 pounds

Metric system

1 gram = 1,000 milligrams

1 kilogram = 1,000 grams

Liquid

Imperial system

1 pint = 16 ounces

1 quart = 2 pints

1 gallon = 4 quarts

Metric system

1 liter = 1,000 milliliters

Time

1 minute = 60 seconds

1 hour = 60 minutes

1 day = 24 hours

1 year = 365 days

To change from one unit to another, we can divide either side of a conversion relationship by the other and always get 1.

For example:

$$1 \text{ yard} = 3 \text{ feet} \Rightarrow \frac{1 \text{ yard}}{3 \text{ feet}} = 1 \text{ or } 1 \text{ gallon} = 4 \text{ quarts} \Rightarrow \frac{4 \text{ quarts}}{1 \text{ gallon}} = 1.$$

Since we can multiply any quantity by one, we can easily change measurements from one unit to another. So to convert 8 feet to inches, we multiply 8 feet by $\frac{12 \text{ inches}}{1 \text{ feet}}$. The feet cancel out to get 8(12) = 96 inches.

PROBLEMS

(A) Convert 15 kilometers to meters.

(B) Convert 6 pints to gallons.

(C) Convert 80 years to seconds.

SOLUTIONS

(A) $15 \text{ kilometers} \cdot \dfrac{1{,}000 \text{ meters}}{1 \text{ kilometers}} = 15{,}000 \text{ meters}$

(B) $6 \text{ pints} \cdot \dfrac{1 \text{ quart}}{2 \text{ pints}} \cdot \dfrac{1 \text{ gallon}}{2 \text{ quarts}} = \dfrac{3}{4} \text{ gallon}$

(C) $80 \text{ years} \cdot \dfrac{365 \text{ days}}{1 \text{ year}} \cdot \dfrac{24 \text{ hours}}{1 \text{ day}} \cdot \dfrac{60 \text{ minutes}}{1 \text{ hour}} \cdot \dfrac{60 \text{ seconds}}{1 \text{ minute}} = 2{,}522{,}880{,}000 \text{ seconds}$

PROBLEM

If 1 kilogram = 2.2 pounds, convert 180 pounds to kilograms.

SOLUTION

$180 \text{ pounds} \cdot \dfrac{1 \text{ kilogram}}{2.2 \text{ pounds}} = 81.8 \text{ kilograms}$

PROBLEM

The speed limit for a Canadian road is 80 kilometers per hour. If a car is traveling at 60 mph, is it speeding? (1 mile = 1.6 kilometer).

SOLUTION

$\dfrac{60 \text{ miles}}{\text{hour}} \cdot \dfrac{1.6 \text{ kilometers}}{1 \text{ mile}} = 96 \dfrac{\text{kilometers}}{\text{hour}}$ so the car is speeding.

PROBLEM

Convert 12 feet per second to miles per hour.

SOLUTION

$$\frac{12\ \cancel{\text{feet}}}{\cancel{\text{second}}}\cdot\frac{1\ \text{mile}}{5,280\ \cancel{\text{feet}}}\cdot\frac{60\ \cancel{\text{seconds}}}{1\ \cancel{\text{minute}}}\cdot\frac{60\ \cancel{\text{minutes}}}{1\ \text{hour}}=8.18\frac{\text{miles}}{\text{hour}}$$

Drill Questions

1. If c is any odd integer, which one of the following must be an even integer?

   (A) $\frac{c}{2}$
   (B) $3c - 1$
   (C) $c^2 + 2$
   (D) $2c + 1$

2. What is the value of $|7 - 13| - |-2 - 9|$?

   (A) 16
   (B) 5
   (C) -5
   (D) -16

3. Which one of the following is an irrational number whose value lies between 7 and 8?

   (A) $\sqrt[3]{400}$
   (B) $7.\overline{2}$
   (C) $\sqrt{65}$
   (D) $\frac{25}{3}$

4. Which one of the following numbers is a prime number between 90 and 100?

   (A) 91
   (B) 94
   (C) 95
   (D) 97

5. Which one of the following inequalities describes a graph on the number line that includes all real numbers?

   (A) $x \geq 1$ and $x \leq 4$
   (B) $x \geq 1$ or $x \leq 4$
   (C) $x \leq 1$ or $x \geq 4$
   (D) $x \leq 1$ and $x \geq 4$

6. If 15 is a factor of x, which one of the following is true for any x?

   (A) 30 must be a factor of x.
   (B) Each of 3 and 5 must be factors of x.
   (C) x must be a prime number greater than 15.
   (D) The only prime factors of x are 3 and 5.

7. The number $0.\overline{4}$ is equivalent to which fraction?

   (A) $\frac{4}{5}$
   (B) $\frac{4}{7}$
   (C) $\frac{4}{9}$
   (D) $\frac{4}{11}$

8. The number 0.00000482 is equivalent to

   (A) 4.82×10^{-5}
   (B) 4.82×10^{-6}
   (C) 4.82×10^{-7}
   (D) $\frac{4.82}{10^{-7}}$

9. Convert 10 miles per hour to meters per second if 1 meter = 3.28 feet.

   (A) 4.47 meters/sec
   (B) 22.36 meters/sec
   (C) 48.11 meters/sec
   (D) 268.28 meters/sec

10. Which one of the following has no solution for x?

   (A) $|x| > 0$
   (B) $|x| = 0$
   (C) $|x| = -x$
   (D) $|x| < 0$

Answers to Drill Questions

1. **(B)** The product of 3 and an odd integer must be an odd integer. The difference of an odd integer and 1 must be an even integer. For example, suppose $c = 5$. Then $3c - 1 = (3)(5) - 1 = 15 - 1 = 14$, which is an even integer.

2. **(C)** Recall that the absolute value of any quantity must be nonnegative. Then $|7-13| = |-6| = 6$ and $|-2-9| = |-11| = 11$. Then $6-11 = 6+(-11) = -5$.

3. **(A)** $\sqrt[3]{400}$ is irrational because we cannot find any integer whose cube is exactly 400. Note that since $7^3=343$ and $8^3= 512$, we know that $\sqrt[3]{400}$ has a value between 7 and 8. (It is approximately 7.368.)

4. **(D)** A prime number can be divided by only two numbers, itself and 1. The number 97 is prime because it is divisble by only 1 and 97.

5. **(B)** The inequality $x \geq 1$ includes all numbers greater than or equal to 1. The inequality $x \leq 4$ includes all numbers less than or equal to 4. Every number must satisfy at least one of these conditions. (Some numbers, such as 2, satisfy both conditions.) Thus, the graph of $x \geq 1$ or $x \leq 4$ includes all numbers on the number line.

6. **(B)** If 15 is a factor of x, then x must be divisible by 15 and by any factor of 15. Each of 3 and 5 is a factor of 15. Let $x = 30$. Note that 15 is a factor of 30, since $30 \div 15 = 2$. In addition, we note that each of 3 and 5 is a factor of 30, since $30 \div 3 = 10$ and $30 \div 5 = 6$.

7. **(C)** Let $N = 0.\overline{4}$, so that $10N = 4.\overline{4}$. By subtracting the first equation from the second equation, we get $9N = 4$. Thus, $N = \frac{4}{9}$. Note that we can check this answer by long division. (Divide 9 into 4 to get $0.\overline{4}$.)

8. **(B)** Placing the carat after the 4, work backwords 6 places to the true decimal point so $0.00000482 = 4.82 \times 10^{-6}$.

9. **(A)** $$\frac{10 \cancel{\text{miles}}}{\cancel{\text{hour}}} \cdot \frac{5280 \cancel{\text{feet}}}{1 \cancel{\text{mile}}} \cdot \frac{1 \text{ meters}}{3.28 \cancel{\text{feet}}} \cdot \frac{1 \cancel{\text{hour}}}{60 \cancel{\text{minutes}}} \cdot \frac{1 \cancel{\text{minute}}}{60 \text{ seconds}} = 4.47 \frac{\text{meters}}{\text{second}}.$$

10. **(D)** The absolute value of any number must be nonnegative. This means that $|x|$ must be greater than or equal to zero. Thus, $|x| < 0$ has no solution.

CHAPTER 3
Algebra and Functions

CHAPTER 3

ALGEBRA AND FUNCTIONS

EQUATIONS

An **equation** is defined as a statement that two separate expressions are equal. A **solution** to an equation containing a single variable is a number that makes the equation true when it is substituted for the variable.

For example, in the equation $3x = 18$, 6 is the solution since $3(6) = 18$. Depending on the equation, there can be more than one solution. Equations with the same solutions are said to be **equivalent equations**. An equation without a solution is said to have a solution set that is the **empty** or **null** set, represented by ϕ.

Replacing an expression within an equation by an equivalent expression will result in a new equation with solutions equivalent to the original equation. For example, suppose we are given the equation

$$3x + y + x + 2y = 15.$$

By combining like terms, we get

$$3x + y + x + 2y = 4x + 3y.$$

Since these two expressions are equivalent, we can substitute the simpler form into the equation to get

$$4x + 3y = 15$$

Performing the same operation to both sides of an equation by the same expression will result in a new equation that is equivalent to the original equation.

ADDITION OR SUBTRACTION

$$y + 6 = 10$$

We can add (-6) to both sides:

$$y + 6 + (-6) = 10 + (-6)$$

$$y + 0 = 10 - 6 = 4$$

MULTIPLICATION OR DIVISION

$$3x = 6$$

We can divide both sides by 3:

$$\frac{3x}{3} = \frac{6}{3}$$

$$x = 2$$

So $3x = 6$ is equivalent to $x = 2$.

PROBLEM

Solve for x, justifying each step.

$$3x - 8 = 7x + 8$$

SOLUTION

$3x - 8 = 7x + 8$	
$3x - 8 + 8 = 7x + 8 + 8$	Add 8 to both sides
$3x + 0 = 7x + 16$	Additive inverse property
$3x = 7x + 16$	Additive identity property
$3x - 7x = 7x + 16 - 7x$	Add $(-7x)$ to both sides

$-4x = 7x - 7x + 16$ Commutative property

$-4x = 0 + 16$ Additive inverse property

$-4x = 16$ Additive identity property

$\frac{-4x}{-4} = \frac{16}{-4}$ Divide both sides by -4

$x = -4$

CHECK YOUR WORK!

Replacing x with -4 in the original equation:

$$3x - 8 = 7x + 8$$

$$3(-4) - 8 = 7(-4) + 8$$

$$-12 - 8 = -28 + 8$$

$$-20 = -20$$

LINEAR EQUATIONS

A **linear equation** with one unknown is one that can be put into the form $ax + b = 0$, where a and b are constants, and $a \neq 0$.

To solve a linear equation means to transform it into the form $x = \frac{-b}{a}$.

A. If the equation has unknowns on both sides of the equality, it is convenient to put similar terms on the same sides. Refer to the following example:

$4x + 3 = 2x + 9$

$4x + 3 - 2x = 2x + 9 - 2x$ Add $-2x$ to both sides

$(4x - 2x) + 3 = (2x - 2x) + 9$ Commutative property

$2x + 3 = 0 + 9$ Additive inverse property

$2x + 3 - 3 = 0 + 9 - 3$	Add -3 to both sides
$2x = 6$	Additive inverse property
$\frac{2x}{2} = \frac{6}{2}$	Divide both sides by 2
$x = 3$	

B. If the equation appears in fractional form, it is necessary to transform it using cross-multiplication, and then repeat the same procedure as in (A). For example,

$$\frac{3x + 4}{3} = \frac{7x + 2}{5}$$

Cross-multiply as follows:

$$\frac{3x + 4}{3} \times \frac{7x + 2}{5}$$

to obtain:

$$3(7x + 2) = 5(3x + 4).$$

This is equivalent to:

$$21x + 6 = 15x + 20,$$

which can be solved as in (A).

$21x + 6 = 15x + 20$	
$21x - 15x + 6 = 15x - 15x + 20$	Add $-15x$ to both sides
$6x + 6 - 6 = 20 - 6$	Combine like terms and add -6 to both sides
$6x = 14$	Combine like terms
$\frac{6x}{6} = \frac{14}{6}$	Divide both sides by 6
$x = \frac{7}{3}$	

FACTOR THEOREM

> If $x = c$ is a solution of the equation $f(x) = 0$, then $(x - c)$ is a **factor** of $f(x)$.

Example:

Let $f(x) = 2x^2 - 5x - 3$. By inspection, we can determine that $2(3)^2 - (5)(3) - 3 = (2)(9) - (5)(3) - 3 = 0$. In this example, $c = 3$, so $(x - 3)$ is also a factor of $2x^2 - 5x - 3$.

Example:

Let $f(x) = x^3 + 3x^2 - 4$. By inspection, we can determine that $(-2)^3 + 3(-2)^2 - 4 = -8 + (3)(4) - 4 = 0$. In this example, $c = -2$, so $(x + 2)$ is also a factor of $x^3 + 3x^2 - 4$.

REMAINDER THEOREM

> If a is any constant and if the polynomial $p(x)$ is divided by $(x - a)$, the **remainder** is $p(a)$.

Example:

Given a polynomial $p(x) = 2x^3 - x^2 + x + 4$, divided by $x - 1$, the remainder is $P(1) = 2(1)^3 - (1)^2 + 1 + 4 = 6$.

That is, $2x^3 - x^2 + x + 4 = q(x) + \dfrac{6}{(x - 1)}$, where $q(x)$ is a polynomial.

Note that in this case $a = 1$.

Also, by using long division, we get $q(x) = 2x^2 + x + 2$.

Example:

Given a polynomial $p(x) = x^4 + x - 50$ divided by $x + 3$, the remainder is $p(-3) = 28$.

That is, $x^4 + x - 50 = q(x) + \frac{28}{(x+3)}$, where $q(x)$ is a polynomial.

Note that in this case $a = -3$.

Also, by using long division, we get $q(x) = x^3 - 3x^2 + 9x - 26$.

SIMULTANEOUS LINEAR EQUATIONS

Two or more equations of the form $ax + by = c$, where a, b, c are constants and $a, b \neq 0$ are called **linear equations** with two unknown variables, or **simultaneous equations.**

Equations with more than one unknown variable are solvable only if you have as many equations as unknown variables.

There are several ways to solve systems of linear equations with two variables. Three of the basic methods are:

Method 1: **Substitution**—Find the value of one unknown in terms of the other. Substitute this value in the other equation and solve.

Method 2: **Addition or subtraction**—If necessary, multiply the equations by numbers that will make the coefficients of one unknown in the resulting equations numerically equal. If the signs of equal coefficients are the same, subtract the equations; otherwise add. The result is one equation with one unknown; we solve it and substitute the value into the other equations to find the unknown that we first eliminated.

Method 3: **Graph**—Graph both equations. The point of intersection of the drawn lines is a simultaneous solution for the equations, and its coordinates correspond to the answer that would be found by substitution or addition/subtraction.

A system of linear equations is **consistent** if there is only one solution for the system.

A system of linear equations is **inconsistent** if it does not have any solutions.

Inconsistent equations represent parallel lines, which are discussed later in this chapter.

PROBLEM

Solve the system of equations.

1. $x + y = 3$
2. $3x - 2y = 14$

SOLUTION

Method 1 (Substitution): From equation (1), we get $y = 3 - x$. Substitute this value into equation (2) to get

$$\begin{aligned} 3x - 2(3 - x) &= 14 \\ 3x - 6 + 2x &= 14 \\ 5x &= 20 \\ x &= 4 \end{aligned}$$

Substitute $x = 4$ into either of the original equations to find $y = -1$.

The answer is $x = 4, y = -1$.

Method 2 (Addition or subtraction): If we multiply equation (1) by 2 and add the result to equation (2), we get

$$\begin{array}{r} 2x + 2y = 6 \\ +\ 3x - 2y = 14 \\ \hline 5x + 0 = 20 \\ x = 4 \end{array}$$

Then, as in Method 1, substitute $x = 4$ into either of the original equations to find $y = -1$.

The answer is $x = 4, y = -1$.

Method 3 (Graphing): Find the point of intersection of the graphs of the equations. To graph these linear equations, solve for y in terms of x. The equations will be in the form $y = mx + b$, where m is the slope and b is the intercept on the y-axis. This is the **slope-intercept form** of the equation.

$$x + y = 3$$

Subtract x from both sides:

$$y = 3 - x$$

$$3x - 2y = 14$$

Subtract $3x$ from both sides:

$$-2y = 14 - 3x$$

Divide by -2:

$$y = -7 + \frac{3}{2}x$$

The graph of each of the linear functions can be determined by plotting only two points. For example, for $y = 3 - x$, let $x = 0$, then $y = 3$. Let $x = 1$, then $y = 2$. The two points on this first line are (0, 3) and (1, 2). For $y = -7 + \frac{3}{2}x$, let $x = 0$, then $y = -7$. Let $x = 2$, then $y = -4$. The two points on this second line are $(0, -7)$ and $(2, -4)$.

To find the point of intersection P of the two lines, graph them.

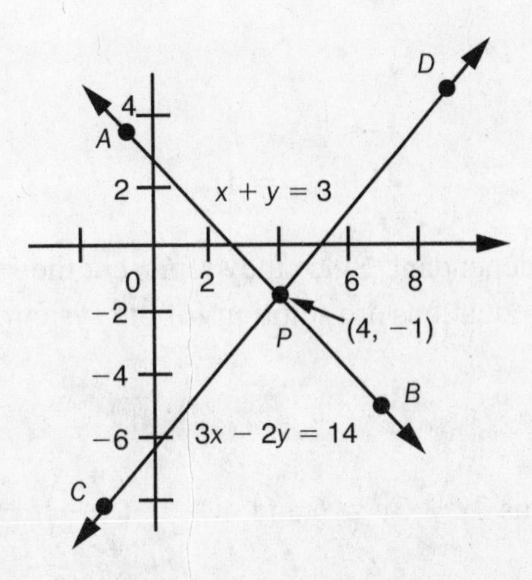

AB is the graph of equation (1), and *CD* is the graph of equation (2). The point of intersection *P* of the two graphs is the only point on both lines. The coordinates of *P* satisfy both equations and represent the desired solution of the problem. From the graph, *P* seems to be point (4, −1). These coordinates satisfy both equations, and hence are the exact coordinates of the point of intersection of the two lines.

CHECK YOUR WORK!

To show that (4, −1) satisfies both equations, substitute this point into both equations.

$x + y = 3$	$3x - 2y = 14$
$4 + (-1) = 3$	$3(4) - 2(-1) = 14$
$4 - 1 = 3$	$12 + 2 = 14$
$3 = 3$	$14 = 14$

DEPENDENT EQUATIONS

Dependent equations are equations that represent the same line; therefore, every point on the line of a dependent equation represents a solution.

Since there are an infinite number of points on a line, there are an infinite number of simultaneous solutions.

Example:

$$2x + y = 8$$
$$4x + 2y = 16$$

These equations are dependent. Since they represent the same line, all points that satisfy either of the equations are solutions of the system.

PROBLEM

Solve the equations $2x + 3y = 6$ and $y = -\left(\frac{2x}{3}\right) + 2$ simultaneously.

SOLUTION

We have two equations and two unknowns:

$$2x + 3y = 6$$

and

$$y = -\left(\frac{2x}{3}\right) + 2$$

As with all simultaneous equations, there are several methods of solution. Since equation (2) already gives us an expression for y, we use the method of substitution.

Substitute $-\left(\frac{2x}{3}\right) + 2$ for y in equation (1):

$$2x + 3\left(-\frac{2x}{3} + 2\right) = 6$$

Distribute:

$$2x - 2x + 6 = 6$$

$$6 = 6$$

Although the result $6 = 6$ is true, it indicates no single solution for x. No matter what real number x is, if y is determined by equation (1), then equation (1) will always be satisfied.

The reason for this peculiarity may be seen if we take a closer look at the equation $y = -\left(\frac{2x}{3}\right) + 2$. It is equivalent to $3y = -2x + 6$, or $2x + 3y = 6$.

In other words, the two equations are equivalent. Any pair of values of x and y that satisfies one satisfies the other.

It is hardly necessary to verify that in this case the graphs of the given equations are identical lines, and that there are an infinite number of simultaneous solutions to these equations.

PARALLEL LINES

> Given two linear equations in x, y, their graphs are **parallel** lines if their slopes are equal. If the lines are parallel, they have no simultaneous solution.

Example:

Line $l_1 : 2x - 7y = 14$, Line $l_2 : 2x - 7y = 56$

In the slope-intercept form, the equation for l_1 is $y = \frac{2}{7}x - 2$ and the equation for l_2 is $y = \frac{2}{7}x - 8$. Each line has a slope of $\frac{2}{7}$.

PROBLEM

Solve the equations $2x + 3y = 6$ and $4x + 6y = 7$ simultaneously.

SOLUTION

We have two equations and two unknowns:

$$2x + 3y = 6$$

and

$$4x + 6y = 7$$

Again, there are several methods to solve this problem. We have chosen to multiply each equation by a different number so that when the two equations are added, one of the variables drops out. Thus,

Multiply equation (1) by 2:	$4x + 6y = 12$	(3)
Multiply equation (2) by -1:	$+\ -4x - 6y = -7$	(4)
Add equations (3) and (4):	$0 = 5$	

We obtain a peculiar result!

Actually, what we have shown in this case is that there is no simultaneous solution to the given equations because $0 \neq 5$. Therefore, there is no simultaneous solution to these two equations, and hence no point satisfying both.

The straight lines that are the graphs of these equations must be parallel if they never intersect, but not identical, which can be seen from the graph of these equations.

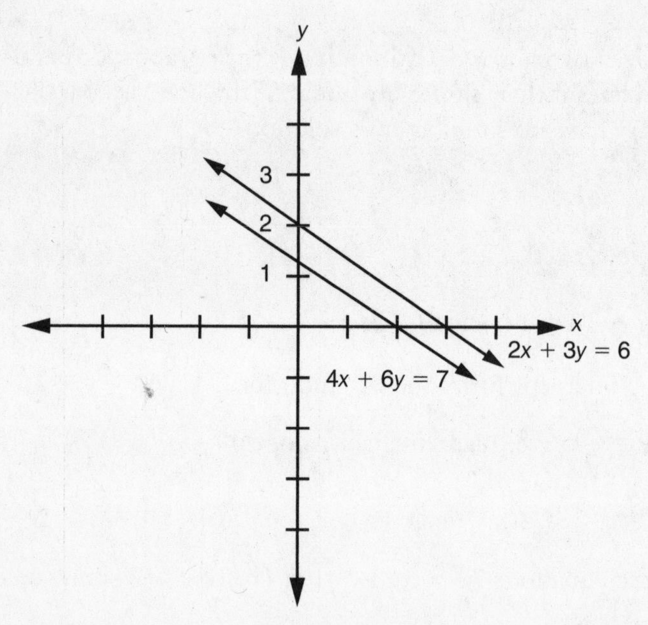

PERPENDICULAR LINES

> If the slopes of the graphs of two lines are negative reciprocals of each other, the lines are **perpendicular** to each other.

An example of two numbers that are negative reciprocals of each other are 2 and $-\frac{1}{2}$. (Remember: $2 = \frac{2}{1}$.)

Example:

l_3: $5x + 6y = 30$, l_4: $6x - 5y = 90$

In the slope-intercept form, the equation for l_3 is $y = -\frac{5}{6}x + 5$ and the equation for l_4 is $y = \frac{6}{5}x - 18$. The slope of l_3, which is $-\frac{5}{6}$, is the negative reciprocal of the slope of l_4, which is $\frac{6}{5}$. Therefore, l_3 is perpendicular to l_4.

To summarize:

- Parallel lines have slopes that are equal.
- Perpendicular lines have slopes that are negative reciprocals of each other.

ABSOLUTE VALUE EQUATIONS

The **absolute value** of a, denoted $|a|$, is defined as

$|a| = a$ when $a > 0$,

$|a| = -a$ when $a < 0$,

$|a| = 0$ when $a = 0$.

When the definition of absolute value is applied to an equation, the quantity within the absolute value symbol may have two values. This value can be either positive or negative before the absolute value is taken. As a result, each absolute value equation actually contains two separate equations.

When evaluating equations containing absolute values, proceed as follows:

$|5 - 3x| = 7$ is valid if either

$$5 - 3x = 7 \quad \text{or} \quad 5 - 3x = -7$$

$$-3x = 2 \qquad\qquad -3x = -12$$

$$x = -\frac{2}{3} \qquad\qquad x = 4$$

The solution set is therefore $x = \left(-\frac{2}{3}, 4\right)$

Remember, the absolute value of a number cannot be negative. So the equation $|5x + 4| = -3$ would have no solution.

INEQUALITIES

The solution of a given inequality in one variable x consists of all values of x for which the inequality is true.

> A **conditional inequality** is an inequality whose validity depends on the values of the variables in the sentence. That is, certain values of the variables will make the sentence true, and others will make it false.

The sentence $3 - y > 3 + y$ is a conditional inequality for the set of real numbers, since it is true for any replacement less than 0 and false for all others, or $y < 0$ is the solution set.

> An **absolute inequality** for the set of real numbers means that for *any* real value for the variable, x, the sentence is always true.

The sentence $x + 5 > x + 2$ is an absolute inequality because the expression on the left is greater than the expression on the right.

A sentence is **inconsistent** if it is always false when its variables assume allowable values.

The sentence $x + 10 < x + 5$ is inconsistent because the expression on the left side is always greater than the expression on the right side.

The sentence $5y < 2y + y$ is inconsistent for the set of non-negative real numbers. For any y greater than 0, the sentence is always false.

Two inequalities are said to have the same **sense** if their signs of inequality point in the same direction.

The sense of an inequality remains the same if both sides are multiplied or divided by the same *positive* real number.

Example:

For the inequality $4 > 3$, if we multiply both sides by 5, we will obtain:

$$4 \times 5 > 3 \times 5$$

$$20 > 15$$

The sense of the inequality does not change.

If each side of an inequality is multiplied or divided by the same *negative* real number, however, the sense of an inequality becomes opposite.

Example:

For the inequality $4 > 3$, if we multiply both sides by -5, we would obtain:

$$4 \times (-5) < 3 \times (-5)$$
$$-20 < -15.$$

The sense of the inequality becomes opposite.

If $a > b$ and a, b, and n are positive real numbers, then

$$a^n > b^n \text{ and } a^{-n} < b^{-n}$$

If $x > y$ and $q > p$, then $x + q > y + p$.

If $x > y > 0$ and $q > p > 0$, then $xq > yp$.

> Inequalities that have the same solution set are called **equivalent inequalities**.

PROBLEM

Solve the inequality $2x + 5 > 9$.

SOLUTION

	$2x + 5 > 9$
Add -5 to both sides:	$2x + 5 + (-5) > 9 + (-5)$
Additive inverse property:	$2x + 0 > 9 + (-5)$
Additive identity property:	$2x > 9 + (-5)$
Combine terms:	$2x > 4$
Multiply both sides by $\frac{1}{2}$ (this is the same as dividing both sides by 2):	$\frac{1}{2}(2x) > \frac{1}{2} \times 4$
	$x > 2$

The solution set is

$X = \{x \mid x > 2\}$

(that is, all x, such that x is greater than 2).

LINEAR INEQUALITIES

Linear inequalities are graphed by shading a section of the coordinate plane. The line is graphed using a solid line if the line is included ($\leq$ or $\geq$) and a dashed line if the line is not included ($<$ or $>$). Shading will be on either side of the line. If the inequality is in the form of $y > mx + b$ or $y \geq mx + b$, the shading is above the line and if the inequality is in the form of $y < mx + b$ or $y \leq mx + b$, the shading is below the line.

PROBLEM

Graph $y \geq 3x - 1$.

SOLUTION

We graph the line $y = 3x - 1$ and make it solid. Since the inequality is $\geq$, we shade above the line.

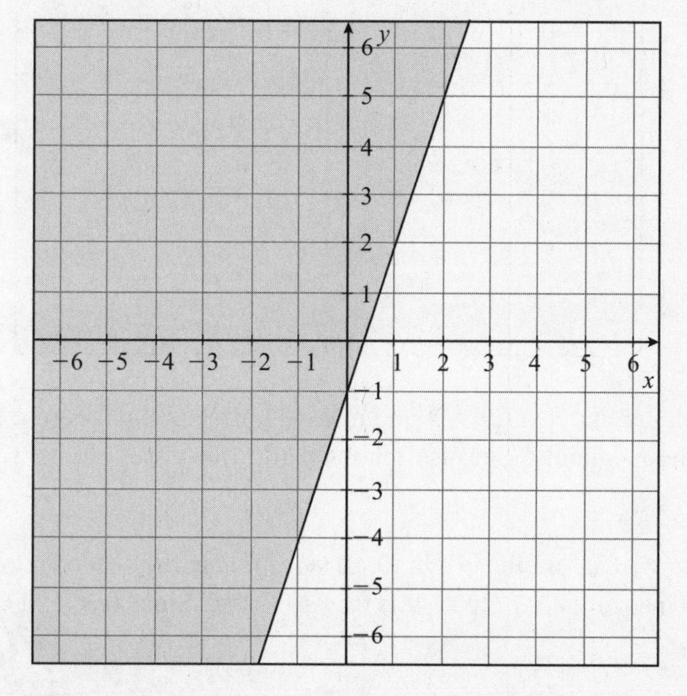

This means every point in the shaded region is true for $y \geq 3x - 1$. Let's check one random point $(-3, 1)$, to show this is correct. We get $1 \geq 3(-3) - 1 = -10$. Is $1 \geq -10$? Yes, so it checks out.

PROBLEM

Graph $x + 2y < 4$.

SOLUTION

$2y < -x + 4$

$y < -\frac{1}{2}x + 2$

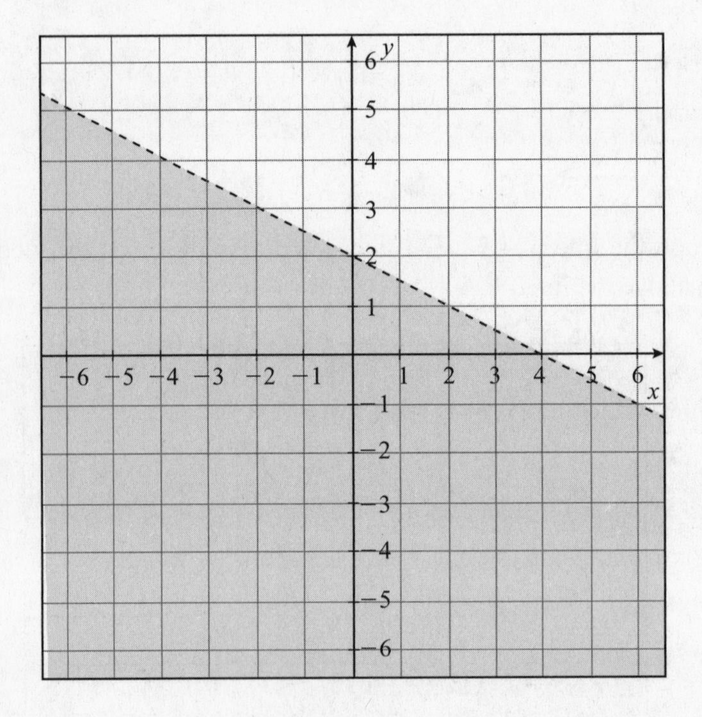

We graph the line $x + 2y = 4$ and make it a dashed line because it is not included in the inequality. Because the inequality for y is < 0, we shade below the line.

To check, we can use the origin (0, 0) since it is in the shaded region. We use the original inequality $x + 2y = 4$. Is $0 + 2(0) < 4$? Since $0 < 4$, it checks out.

Piecewise functions

A **piecewise function** is one where there are two or more rules, based on the value of x.

PROBLEMS

If $f(x)=\begin{cases} x^2-3, & x\geq 0 \\ 2x+1, & x<0 \end{cases}$, find

(A) $f(4)$

(B) $f(3) - f(-1)$

(C) $f(f(0))$

SOLUTIONS

(A) $f(4) = 16 - 3 = 13$

(B) $f(3) - f(-1) = (9 - 3) - (-2 + 1)$
$= 6 - (-1) = 7$

(C) $f(0) = 0 - 3 = -3$

$f(-3) = -6 + 1 = -5$

COMPLEX NUMBERS

As indicated above, real numbers provide the basis for most precalculus mathematics topics. However, on occasion, real numbers by themselves are not enough to explain what is happening. As a result, complex numbers were developed.

> A **complex number** is a number that can be written in the form $a + bi$, where a and b are real numbers and $i = \sqrt{-1}$. The number a is the **real part**, and the number bi is the **imaginary part** of the complex number.

Returning momentarily to real numbers, the square of a real number cannot be negative. More specifically, the square of a positive real number is positive, the square of a negative real number is positive, and the square of 0 is 0.

i is defined to be a number with a property that $i^2 = -1$. Obviously, i is not a real number. C is then used to represent the set of all complex numbers:

$$C = \{a + bi \mid a \text{ and } b \text{ are real numbers}\}.$$

ADDITION, SUBTRACTION, AND MULTIPLICATION OF COMPLEX NUMBERS

Here are the definitions of addition, subtraction, and multiplication of complex numbers.

Suppose $x + yi$ and $z + wi$ are complex numbers. Then (remembering that $i^2 = -1$):

$$(x + yi) + (z + wi) = (x + z) + (y + w)i$$

$$(x + yi) - (z + wi) = (x - z) + (y - w)i$$

$$(x + yi) \times (z + wi) = (xz - wy) + (xw + yz)i$$

PROBLEM

Simplify $(3 + i)(2 + i)$.

SOLUTION

$$\begin{aligned}(3 + i)(2 + i) &= 3(2 + i) + i(2 + i) \\ &= 6 + 3i + 2i + i^2 \\ &= 6 + (3 + 2)i + (-1) \\ &= 5 + 5i\end{aligned}$$

DIVISION OF COMPLEX NUMBERS

Division of two complex numbers is usually accomplished with a special procedure that involves the conjugate of a complex number. The conjugate of $a + bi$ is denoted by $\overline{a + bi}$ and defined by $\overline{a + bi} = a - bi$.

Also,

$$(a + bi)(a - bi) = a^2 + b^2$$

The usual procedure for division is to multiply and divide by the conjugate as shown below. Remember that multiplication and division by the same quantity leaves the original expression unchanged.

$$\frac{x + yi}{z + wi} = \frac{x + yi}{z + wi} \times \frac{z - wi}{z - wi}$$

$$= \frac{(xz + yw) + (-xw + yz)i}{z^2 + w^2}$$

$$= \frac{xz + yw}{z^2 + w^2} + \frac{-xw + yz}{z^2 + w^2}i$$

If a is a real number, then a can be expressed in the form $a = a + 0i$. Hence, every real number is a complex number and $R \subseteq C$.

All the properties of real numbers carry over to complex numbers, so those properties will not be stated again.

QUADRATIC EQUATIONS

Consider the polynomial:

$ax^2 + bx + c = 0$, where $a \neq 0$.

This type of equation is called a **quadratic equation**.

There are several methods to solve quadratic equations, some of which are highlighted here. The first two are based on the fact that if the product of two factors is 0, either one or the other of the factors equals 0. The equation can be solved by setting each factor equal to 0. If you cannot see the factors right away, however, the quadratic formula, which is the last method presented, *always* works.

SOLUTION BY FACTORING

We are looking for two binomials that, when multiplied together, give the quadratic trinomial

$$ax^2 + bx + c = 0$$

This method works easily if $a = 1$ and you can find two numbers whose product equals c and sum equals b. The signs of b and c need to be considered:

- If c is positive, the factors are going to both have the same sign, which is b's sign.
- If c is negative, the factors are going to have opposite signs, with the larger factor having b's sign.

Once you have the two factors, insert them in the general factor format $(x + \_)(x + \_) = 0$.

To solve the quadratic equation, set each factor equal to 0 to yield the solution set for x.

PROBLEM

Solve the quadratic equation $x^2 + 7x + 12 = 0$.

SOLUTION

We need two numbers whose product is $+12$ and sum is $+7$. They would be $+3$ and $+4$, and the quadratic equation would factor to $(x + 3)(x + 4) = 0$.

Therefore, $x + 3 = 0$ or $x + 4 = 0$ would yield the solutions, which are $x = -3, x = -4$.

CHECK YOUR WORK!

Substitute the values into the original quadratic equation:

For $x = -3$, $(-3)^2 + 7(-3) + 12 = 0$, or $9 + (-21) + 12 = 0$. So $x = -3$ is a solution.

Likewise, for $x = -4$, $(-4)^2 + 7(-4) + 12 = 0$, or $16 + (-28) + 12 = 0$. So $x = -4$ is a solution.

PROBLEM

Suppose the quadratic equation is similar to the previous example, but the sign of b is negative. Solve the quadratic equation $x^2 - 7x + 12 = 0$.

SOLUTION

We need two numbers whose product is $+12$ and sum is -7. They would be -3 and -4, and the quadratic would factor to $(x - 3)(x - 4) = 0$.

Therefore, $x - 3 = 0$ or $x - 4 = 0$ would yield the solutions, which are $x = 3, x = 4$.

CHECK YOUR WORK!

Substitute the values into the original quadratic equation:

For $x = 3$, $(3)^2 - 7(3) + 12 = 0$, or $9 - 21 + 12 = 0$. So $x = 3$ is a solution.

Likewise, for $x = 4$, $(4)^2 - 7(4) + 12 = 0$, or $16 - 28 + 12 = 0$. So $x = 4$ is a solution.

PROBLEM

As a final example, solve the quadratic equation $x^2 + 4x - 12 = 0$.

SOLUTION

We need two numbers whose product is -12 and sum is $+4$. They would be $+6$ and -2 (note that the larger numeral gets the $+$ sign, the sign of b). The quadratic equation would factor to $(x + 6)(x - 2) = 0$.

Therefore, $x + 6 = 0$ or $x - 2 = 0$ would yield the solutions, which are $x = -6, x = 2$.

CHECK YOUR WORK!

Substitute the values into the original quadratic equation:

For $x = -6$, $(-6)^2 + 4(-6) - 12 = 0$, or $36 + (-24) - 12 = 0$. So $x = -6$ is a solution.

Likewise, for $x = 2$, $(2)^2 + 4(2) - 12 = 0$, or $4 + 8 - 12 = 0$. So $x = 2$ is a solution.

SUM OF TWO SQUARES

If the quadratic consists of the difference of only two terms of the form $ax^2 - c$, and you can recognize them as perfect squares, the factors are simply the sum and difference of the square roots of the two terms. Note that a is a perfect square, but not necessarily 1, for this method.

PROBLEM

Solve the quadratic equation $x^2 - 16 = 0$.

SOLUTION

This is the difference of two perfect squares, x^2 and 16, whose square roots are x and 4. So the factors are $(x + 4)(x - 4) = 0$, and the solution is $x = \pm 4$.

CHECK YOUR WORK!

For $x = 4$, $(4)^2 - 16 = 16 - 16 = 0$.

For $x = -4$, $(-4)^2 - 16 = 16 - 16 = 0$.

PROBLEM

Solve the quadratic equation $9x^2 - 36 = 0$.

SOLUTION

This is the difference of two perfect squares, $9x^2$ and 36, whose square roots are $3x$ and 6. So the factors are $(3x + 6)(3x - 6) = 0$, then $3x + 6 = 0$ or $3x - 6 = 0$, and the solution is $x = \pm 2$.

CHECK YOUR WORK!

For $x = 2$, $9(2)^2 - 36 = 36 - 36 = 0$.

For $x = -2$, $9(-2)^2 - 36 = 36 - 36 = 0$.

QUADRATIC FORMULA

If the quadratic equation does not have obvious factors, the roots of the equation can always be determined by the **quadratic formula** in terms of the coefficients a, b, and c as shown below:

$$x = \frac{-b \pm \sqrt{b^2 - 4ac}}{2a}$$

where $(b^2 - 4ac)$ is called the **discriminant** of the quadratic equation.

- If the discriminant is less than zero $(b^2 - 4ac < 0)$, the roots are complex numbers, since the discriminant appears under a radical and square roots of negatives are imaginary numbers. A real number added to an imaginary number yields a complex number.
- If the discriminant is equal to zero $(b^2 - 4ac = 0)$, the roots are real and equal.
- If the discriminant is greater than zero $(b^2 - 4ac > 0)$, then the roots are real and unequal. The roots are rational if and only if a and b are rational and $(b^2 - 4ac)$ is a perfect square; otherwise, the roots are irrational.

PROBLEMS

Compute the value of the discriminant and then determine the nature of the roots of each of the following four equations:

1. $4x^2 - 12x + 9 = 0$
2. $3x^2 - 7x - 6 = 0$
3. $5x^2 + 2x - 9 = 0$
4. $x^2 + 3x + 5 = 0$

SOLUTIONS

1. $4x^2 - 12x + 9 = 0$

   Here, a, b, and c are integers:

   $a = 4$, $b = -12$, and $c = 9$.

Therefore,

$b^2 - 4ac = (-12)^2 - 4(4)(9) = 144 - 144 = 0$

Since the discriminant is 0, the roots are rational and equal.

2. $3x^2 - 7x - 6 = 0$

   Here, a, b, and c are integers:

   $a = 3$, $b = -7$, and $c = -6$.

   Therefore,

   $b^2 - 4ac = (-7)^2 - 4(3)(-6) = 49 + 72 = 121 = 11^2$.

   Since the discriminant is a perfect square, the roots are rational and unequal.

3. $5x^2 + 2x - 9 = 0$

   Here, a, b, and c are integers:

   $a = 5$, $b = 2$, and $c = -9$

   Therefore,

   $b^2 - 4ac = 2^2 - 4(5)(-9) = 4 + 180 = 184$.

   Since the discriminant is greater than zero, but not a perfect square, the roots are irrational and unequal.

4. $x^2 + 3x + 5 = 0$

   Here, a, b, and c are integers:

   $a = 1$, $b = 3$, and $c = 5$

   Therefore,

   $b^2 - 4ac = 3^2 - 4(1)(5) = 9 - 20 = -11$

   Since the discriminant is negative, the roots are complex.

PROBLEM

Solve the equation $x^2 - x + 1 = 0$.

SOLUTION

In this equation, $a = 1$, $b = -1$, and $c = 1$. Substitute into the quadratic formula.

$$x = \frac{-(-1) \pm \sqrt{(-1)^2 - 4(1)(1)}}{2(1)}$$

$$= \frac{1 \pm \sqrt{1-4}}{2}$$

$$= \frac{1 \pm \sqrt{-3}}{2}$$

$$= \frac{1 \pm \sqrt{3}i}{2}$$

$$x = \frac{1 + \sqrt{3}i}{2} \text{ or } x = \frac{1 - \sqrt{3}i}{2}$$

ADVANCED ALGEBRAIC THEOREMS

A. Every polynomial equation $f(x) = 0$ of degree greater than zero has at least one root either real or complex. This is known as the **fundamental theorem of algebra**.

B. Every polynomial equation of degree n has exactly n roots.

C. If a polynomial equation $f(x) = 0$ with real coefficients has a root $a + bi$, then the conjugate of this complex number $a - bi$ is also a root of $f(x) = 0$.

D. If $a + \sqrt{b}$ is a root of polynomial equation $f(x) = 0$ with rational coefficients, then $a - \sqrt{b}$ is also a root, where a and b are rational and $\sqrt{b}$ is irrational.

E. If a rational fraction in lowest terms $\frac{b}{c}$ is a root of the equation $a_n x^n + a_{n-1}x^{n-1} + ... + a_1 x + a_0 = 0$, $a_0 \neq 0$, and the a_i are integers, then b is a factor of a_0 and c is a factor of a_n.

F. Any rational roots of the equation $x^n + q_1 x^{n-1} + q_2 x^{n-2} + ... + q_{n-1}x + q_n = 0$ must be integers and factors of q_n. Note that $q_1, q_2, \ldots, q_n$ are integers.

RELATIONS AND GRAPHS

An **ordered pair** is commonly called a point. It is in the form of (x, y) where x and y are real numbers. A **relation** is a set of points.

PROBLEM

In relation $R = \{(0, 0), (1, 0), (0, 1), (1, 1), (2, 0), (2, 1), (1, 2), (0, 2), (2, 2)\}$, list the set of ordered pairs for which the 2nd member is greater than the first member.

SOLUTION

$(0, 1), (1, 2), (0, 2)$

The **domain** of a relation is the set of all first members of the relation. If a member is repeated, it is listed once.

The **range** of a relation is the set of all second members of the relation. If a member is repeated, it is listed once.

PROBLEM

List the domain and range of the relation

$R = \{(-2, -2), (-1, 0), (-1, -2), (-1, 0)\}$

SOLUTION

Domain: $\{-2, -1\}$ Range: $\mathrm{R} = \{-2, 0\}$

FUNCTIONS: DEFINITION AND NOTATION

A **function** is a relation, a set of points (x, y), such that for every x, there is one and only one y. In short, in a function, the x-values cannot repeat while the y-values can. On the CLEP test, almost all of your graphs will come from functions.

PROBLEM

Which of the following relations are functions?

$A = \{(8,0), (-6,-3), (1,-2), (2,-10)\}$

$B = \{(5,1), (7,f), (5,b), (-3,p)\}$

$C = \{(1,5), (-1,-7), (0,5), (1,-1), (-1,-2)\}$

$D = \{(3,4), (5,4), (8,4), (1,4), (12,4), (0,4)\}$

SOLUTION

A and D are functions. B is not a function as the 5 repeats. C is not a function as both the 1 and -1 repeat.

If a graph is given, you can quickly determine whether it is a function by the **vertical line test**. If it is possible for a vertical line to intersect a graph at more than one point, then the graph is not a function.

PROBLEM

Use the vertical line test to determine which of the following graphs are functions.

(A)

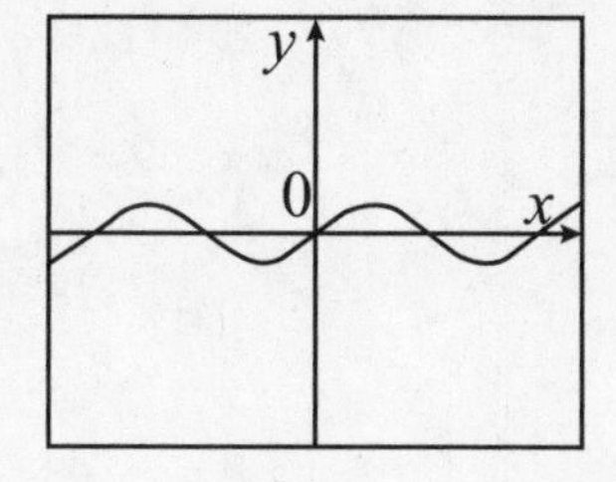

(C)

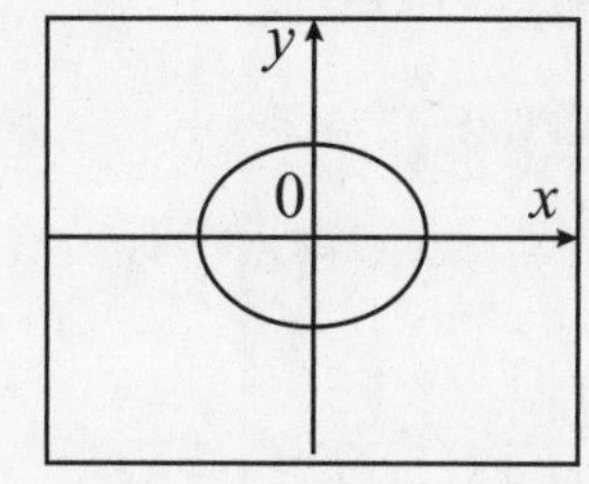

(B)

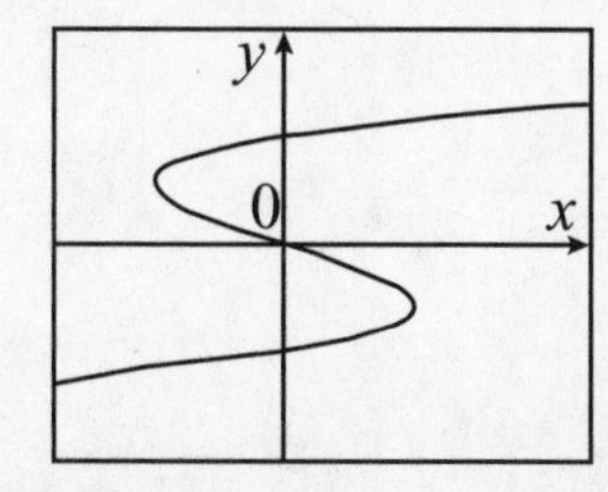

(D)

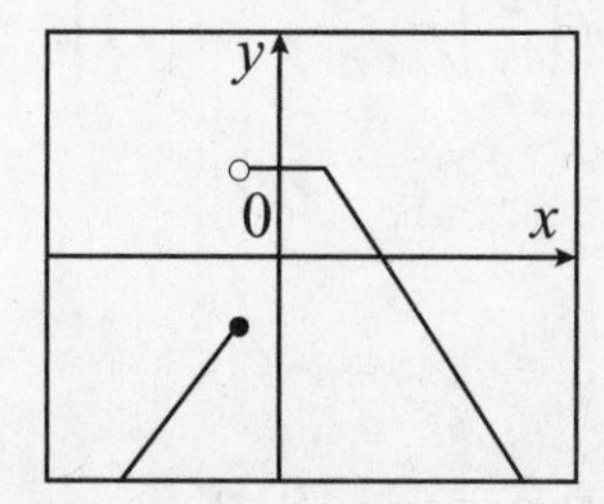

SOLUTION

(A) is a function as any vertical line would only intersect the graph in one location. Choices (B) and (C) are not functions as it is possible to draw a vertical line that will intersect the graph in more than one location. Choice (D) is a function. The open dot indicates the graph does not contain the point while the closed dot indicates the graph does include the point. Thus, a vertical line will intersect the graph in only one location.

Functions can be represented in four different ways. First is the **numerical representation** where functions are given by the set of points: $F = \{(-1, 1), (0, 2), (1, 3), (2, 4), (3, 5)\}$ is an example of a function given numerically. A **graphical representation** gives a picture of the function. In the problem on the previous page, (A) and (D) show functions graphically.

The most common way to represent a function is **symbolically**. We use a special function notation. This notation is either in the form of "$y =$" or "$f(x) =$". In the $f(x)$ notation, we are stating a rule to find y given a value of x. Using this notation, it is easy to **evaluate the function**—plugging in a value of x to find y.

PROBLEMS

If $f(x) = x^2 - 5x + 8$, find

(A) $f(-6)$

(B) $f\left(\frac{3}{2}\right)$

(C) $f(a)$

SOLUTIONS

(A) $f(-6) = (-6)^2 - 5(-6) + 8$

$36 + 30 + 8$

74

(B) $f\left(\frac{3}{2}\right) = \left(\frac{3}{2}\right)^2 - 5\left(\frac{3}{2}\right) + 8$

$\frac{9}{4} - \frac{15}{2} + 8$

$\frac{11}{4}$

(C) $f(a) = a^2 - 5a + 8$

COMPOSITION OF FUNCTIONS

One concept that is tested on the CLEP exam is composition of functions. If we have two functions f and g, we can find $f(g(a))$ or $g(f(a))$, which are different than $f(a) \cdot g(a)$. To find a composition of functions: plug a value into one function, determine an answer, and plug that answer into a second function. $f(g(x))$ can also be written as $(f \text{ o } g)(x)$.

PROBLEMS

If $f(x) = x^2 - x + 1$ and $g(x) = 2x - 1$, find

(A) $f(-1) \cdot g(-1)$

(B) $f(g(-1))$

(C) $g(f(-1))$

(D) $(g \text{ o } f)(x)$

SOLUTIONS

(A) $f(-1)\, g(-1) = (1 + 1 + 1) = (-2 - 1) = 3(-3) = -9$

(B) $g(-1) = 2(-1) - 1 = -3$ so $f(-3) = 9 + 3 + 1 = 13$

(C) $f(-1) = 1 + 1 + 1 = 3$ so $g(3) = 6 - 1 = 5$

(D) $(g \text{ o } f)(x) = g(f(x)) = 2(x^2 - x + 1) + 1 = 2x^2 - 2x + 2 + 1 = 2x^2 - 2x + 2 + 3$

PROBLEM

The graph of $g(x)$ is given in the figure on the next page. Find the value of $g(g(-1))$.

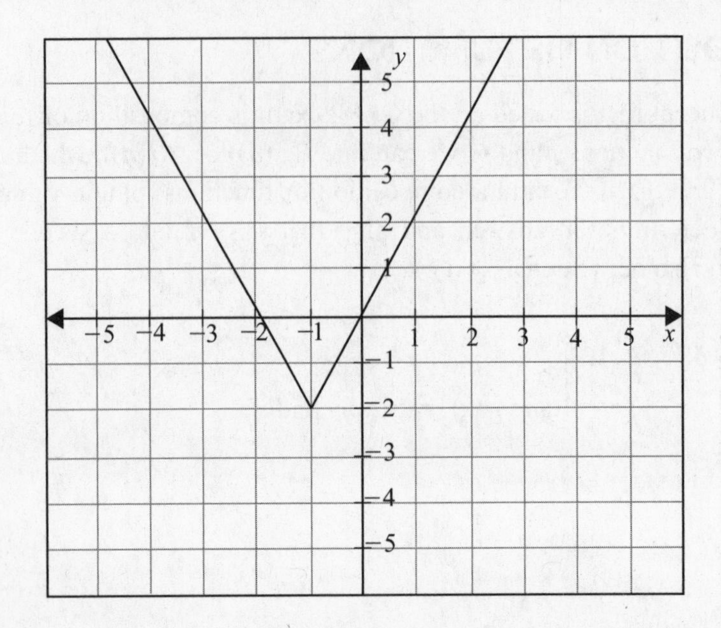

SOLUTION

$g(-1) = -2$. $g(-2) = 0$

LINEAR FUNCTIONS

The type of function which appears most on the CLEP College Mathematics exam are **linear** or **first-degree functions**. The graph of any linear equation is a straight line. The **slope** of a line is a number that measures its steepness. The ratio of the change in y to the change in x is the slope of the line. It is commonly referred to as rise over run.

Slope: Given two points (x_1, y_1) and (x_2, y_2), the slope of the line passing through the points can be written as:

$$m = \frac{\text{rise}}{\text{run}} = \frac{y_2 - y_1}{x_2 - x_1}.$$

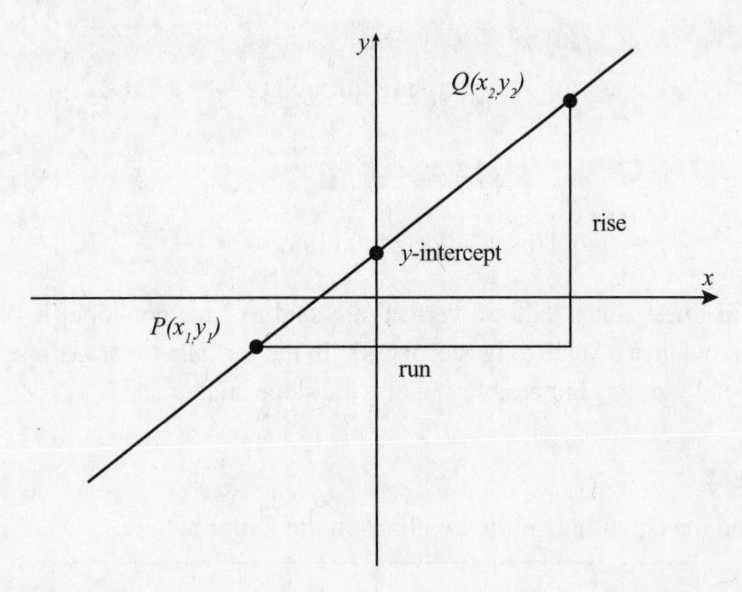

Lines that go up to the right have a positive slope, while lines that go up to the left have a negative slope.

PROBLEMS

Find the slope of the lines passing through the following two points:

(A) (2, 3) and (4, 7)

(B) (−5, 2) and (1, −3)

SOLUTIONS

(A) $m = \frac{7-3}{4-2} = \frac{4}{2} = 2$

(B) $m = \frac{-3-2}{1+5} = \frac{-5}{6}$

HORIZONTAL AND VERTICAL LINES

Horizontal lines: Lines that are horizontal are in the form: y = constant. Horizontal lines have zero slope. In the formula for slope, $y_2 - y_1 = 0$, so the slope is always zero.

PROBLEM

Find the slope of the line passing through (4, –8) and (–2, –8).

SOLUTION

$m = \dfrac{-8+8}{4+2} = \dfrac{0}{6} = 0$ This is a horizontal line.

Vertical lines: Lines that are vertical are said to have no slope (it is so steep that we cannot give a value to its steepness). In the formula for slope, $x_2 - x_1 = 0$, and division by zero is impossible, making the slope undefined.

PROBLEM

Find the equations of the two lines in the figure below.

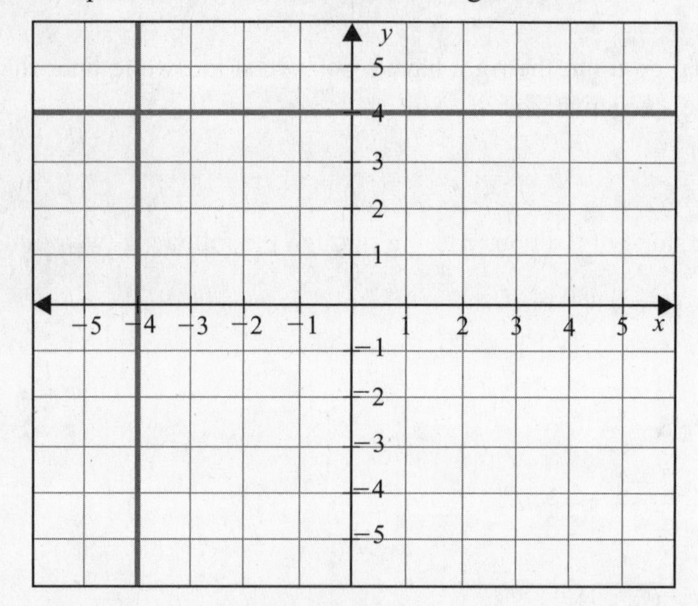

SOLUTION

The horizontal line is $y = 4$. The vertical line is $x = -4$.

LINEAR EQUATIONS

The equations of lines can have several forms, each having its own particular use.

General Form

The general form of an equation is $Ax + By + C = 0$ where A, B, and C are integers. This is the least useful form of a linear equation but it has no fractions in it. An equation in this form often has to be put into the other forms, described below, to get more information about the line, such as its slope. Sometimes answers on the CLEP exam will be placed into general form.

Slope-Intercept Form

All lines except vertical lines will have a ***y*-intercept**, the point in the form of $(0, b)$ at which the line crosses the y-axis. The equation of a line with slope m and y-intercept b is given by $y = mx + b$. If we are given the slope m and the y-intercept b, it is easy to write the equation of a line.

PROBLEM

Find the equation of the line in the figure below.

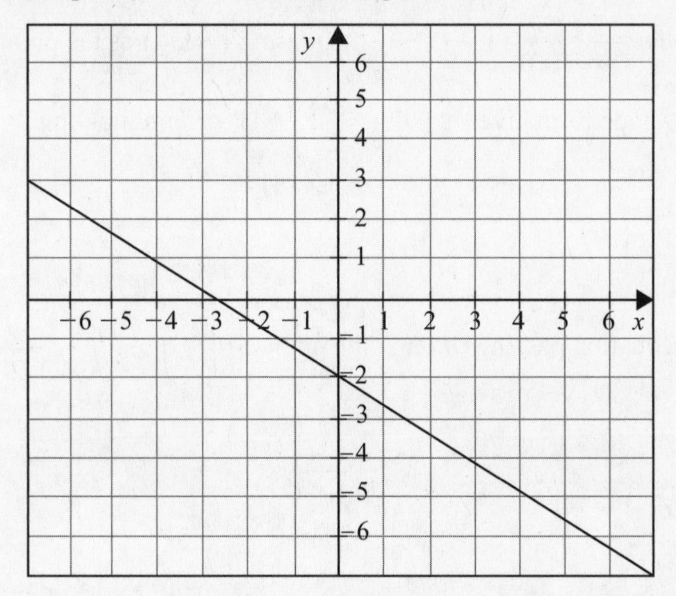

SOLUTION

We see that the y-intercept is $b = -2$. The line goes down 2 units for every 3 units it goes to the right, so the rise is –2 for a run of 3. Therefore, the slope is $m = \frac{-2}{3}$. So the equation is $y = \frac{-2}{3}x - 2$. To transform this equation to general form, clear the equation of fractions by multiplying each term by 3 to get $3y = -2x - 6$ or $2x + 3y + 6 = 0$.

Point-Slope Form

The equation of a line with slope m and passing through the point (x_1, y_1) is given by $y - y_1 = m(x - x_1)$.

If we are given two points, this is the easiest way to find the equation of the line. We need to find the slope first, then plug in either point into the point-slope formula.

PROBLEM

Find the equation passing through the points $(3, -2)$ and $(-1, 4)$.

SOLUTION

The slope $m = \frac{4+2}{-1-3} = \frac{6}{-4} = -\frac{3}{2}$. Choosing $(-1, 4)$ as the point, the point-slope form gives $y - 4 = -\frac{3}{2}(x+1)$ or (multiplying by 2), $2y - 8 = -3x - 3$ General form: $3x + 2y - 5 = 0$.

PROBLEM

Find the equation passing through the points $\left(\frac{3}{4}, 2\right)$ and $\left(\frac{1}{8}, -\frac{1}{2}\right)$.

SOLUTION

The slope $m = \left(\dfrac{2+\frac{1}{2}}{\frac{3}{4}-\frac{1}{8}}\right)$. Multiply the numerator and denominator by the LCD of 8 to clear the fractions. Then $m = \dfrac{16+4}{6-1} = \dfrac{20}{5} = 4$. Choosing $\left(\dfrac{3}{4}, 2\right)$ as the point, $y-2=4\left(x-\dfrac{3}{4}\right)$ or $y-2=4x-3$ which gives $y=4x-1$.

The general form is thus $4x-y-1=0$.

Intercept Form

In intercept form, the equation of a line with x-intercept a and y-intercept b is given by $\dfrac{x}{a}+\dfrac{y}{b}=1$. If we know these intercepts, we can immediately write the equation of the line.

PROBLEM

Find the equation passing through the points (0,4) and $(-3,0)$.

SOLUTION

We can use the intercept form above, $\dfrac{x}{a}+\dfrac{y}{b}=1$, where point (0, 4) says the y-intercept is 4 and (–3,0) says the x-intercept is –3. Since we know the intercepts, we can write the equation directly: $\dfrac{x}{-3}+\dfrac{y}{4}=1$. Clearing the equation of fractions by multiplying through by the LCD = 12 gives $-4x+3y=12$, which in general form is $4x-3y+12=0$.

TRANSFORMATION OF GRAPHS

A curve in the form $y = f(x)$, which is one of the basic common functions, can be transformed in a variety of ways. The shape of the resulting curve stays the same but x- and y-intercepts might change and the graph could be reversed.

The table below describes transformations to a general function $y = f(x)$ with the $f(x) = x^2$ as an example.

Notation	How $f(x)$ changes	Example with $f(x) = x^2$
$f(x) + a$	Moves graph up a units.	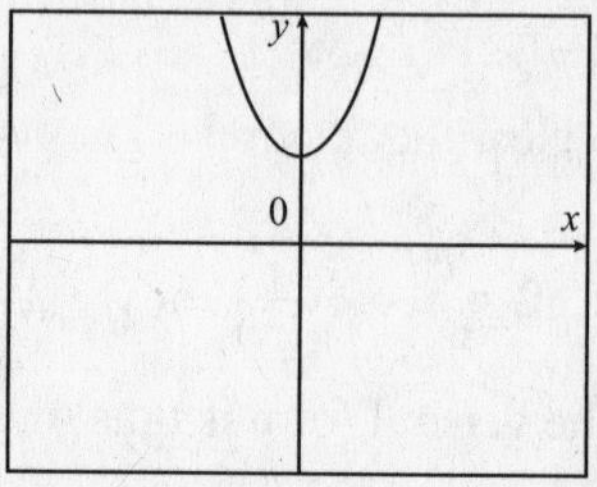
$f(x) - a$	Moves graph down a units.	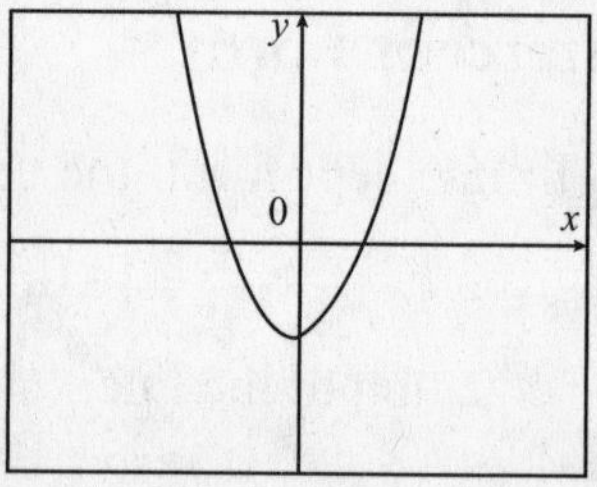
$f(x + a)$	Moves graph a units left.	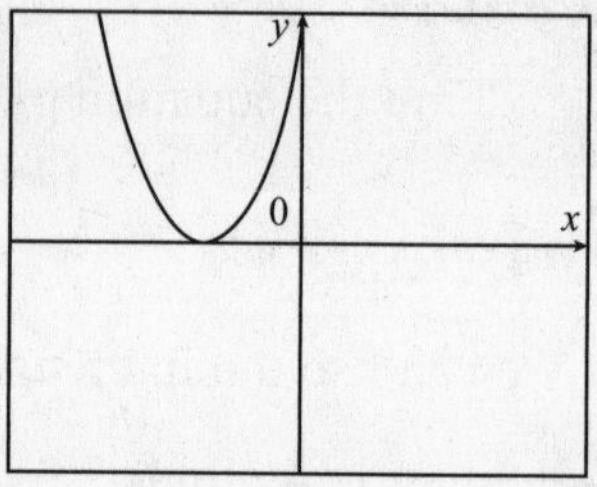
$f(x - a)$	Moves graph a units right.	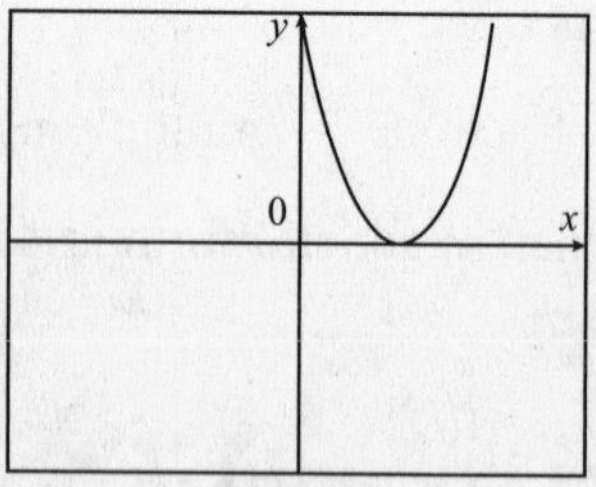
$a \cdot f(x)$	$a > 1$: Vertical stretch	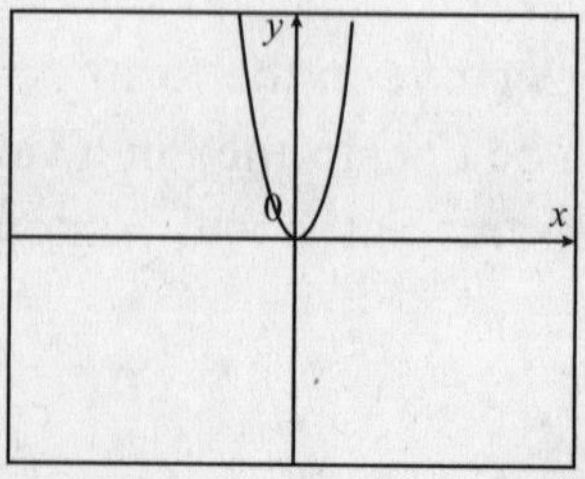

Notation	How $f(x)$ changes	Example with $f(x) = x^2$
$a \cdot f(x)$	$0 < a < 1$: Vertical shrink	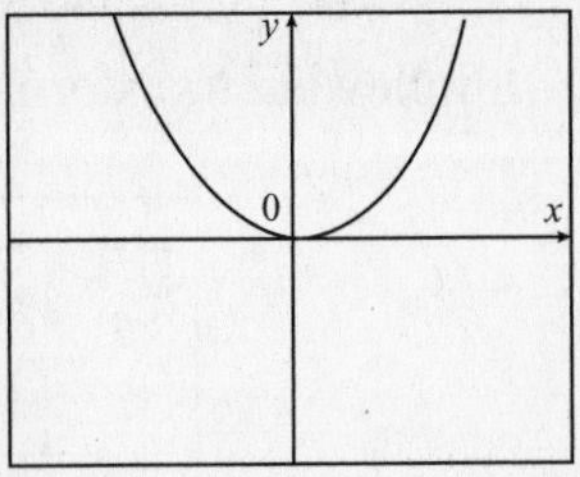
$f(ax)$	$a > 1$: Horizontal compression (for this curve, same effect as vertical stretch)	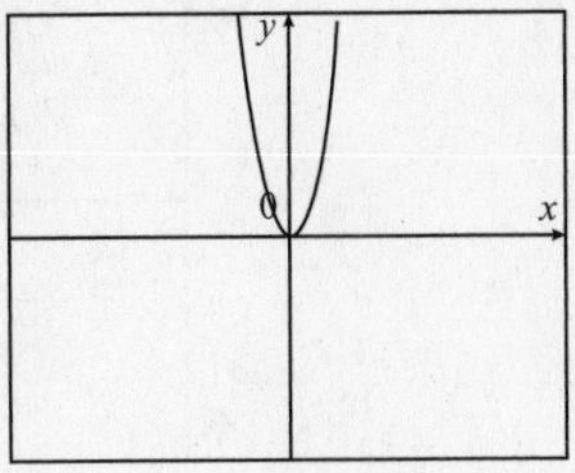
$f(ax)$	$0 < a < 1$: Horizontal elongation (for this curve, same effect as vertical shrink)	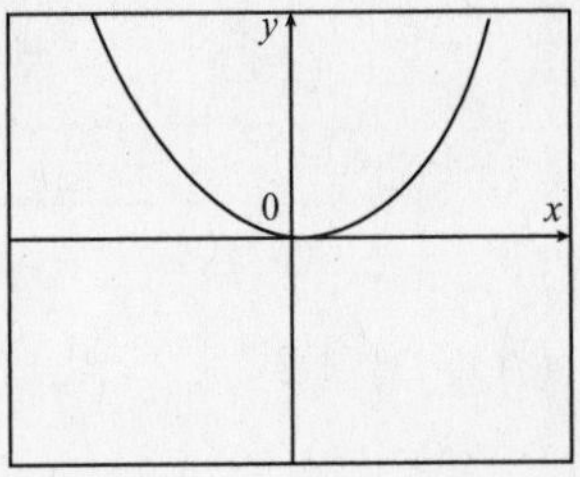
$-f(x)$	Reflection across x-axis	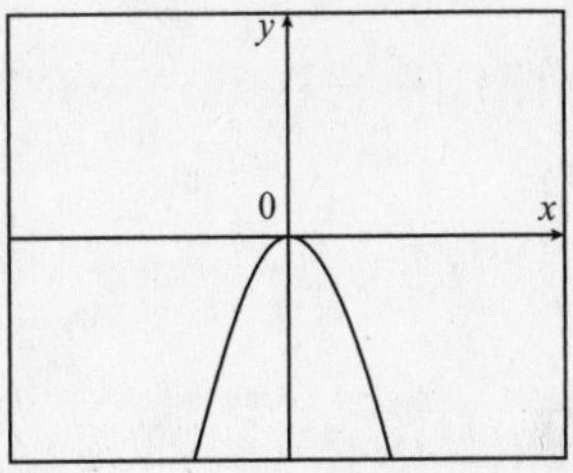
$f(-x)$	Reflection across y-axis	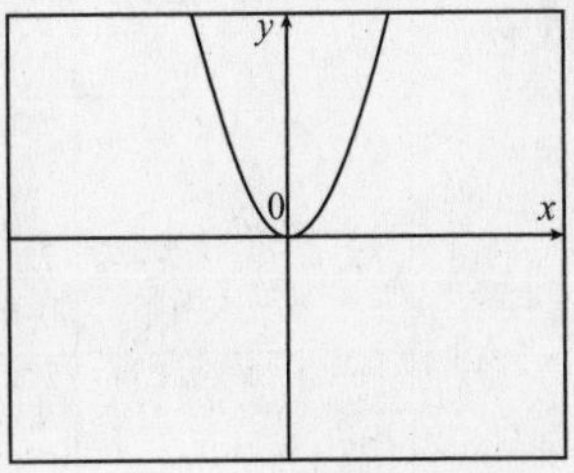

PROBLEMS

Let $g(x)$ be the curve shown in the figure below. Sketch the result of the following transformations.

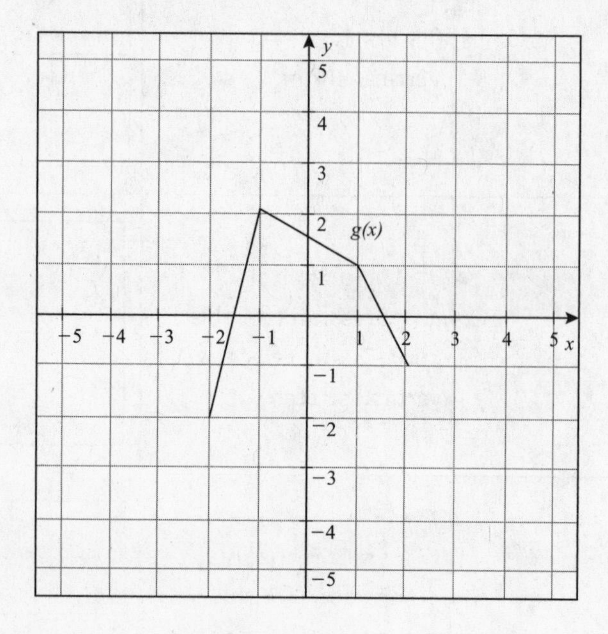

(A) $g(x) + 2$

(B) $g(x) - 1$

(C) $2g(x)$

(D) $-g(x)$

(E) $g(x - 3)$

(F) $g(x + 2) + 1$

(G) $g(2x)$

(H) $g\left(\frac{1}{2}x\right)$

SOLUTIONS

(A) 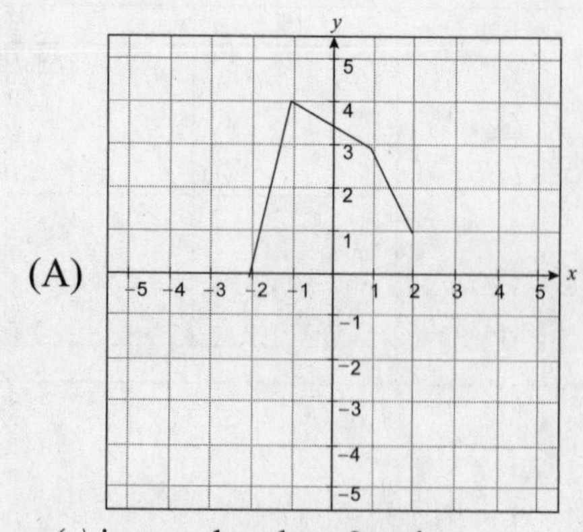

$g(x)$ is translated up 2 units

(B) 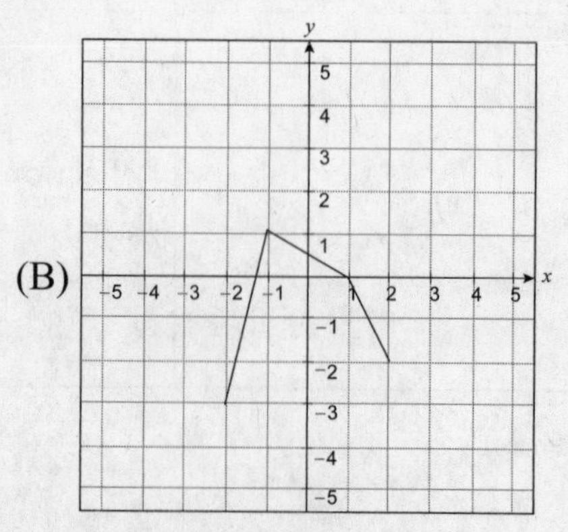

$g(x)$ is translated down 1 unit

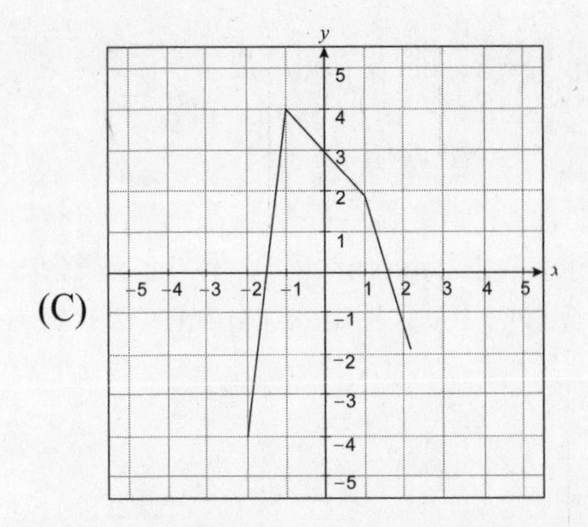

g(x) is vertically stretched 2 units. It goes twice as high and twice as low.

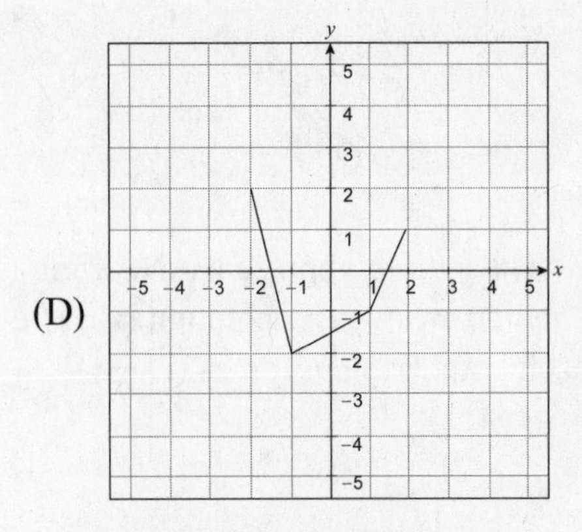

g(x) is reflected across the *x* axis.

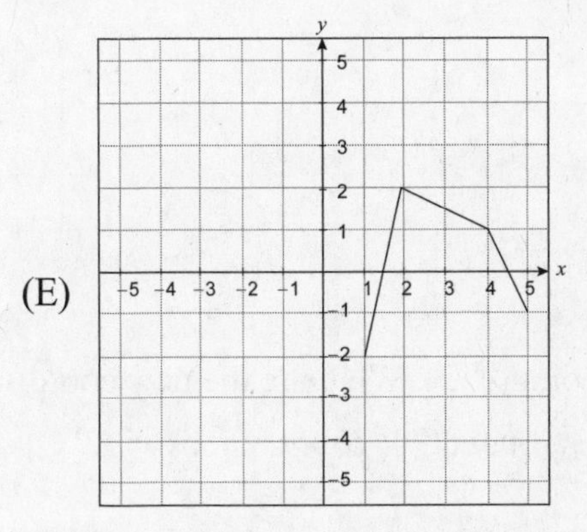

g(x) is translated 3 units to the right.

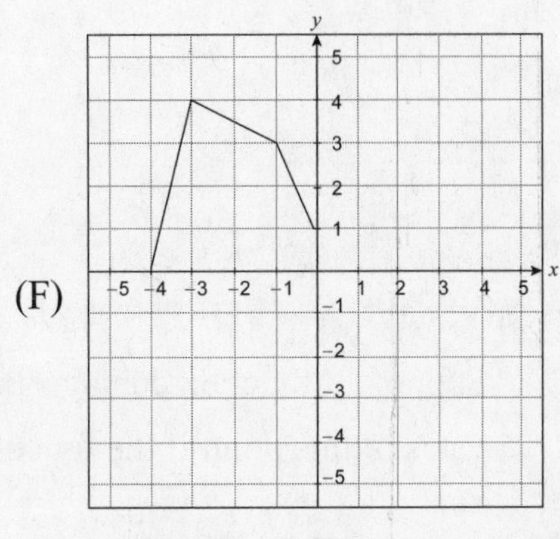

g(x) is translated 2 units to the left and one unit up.

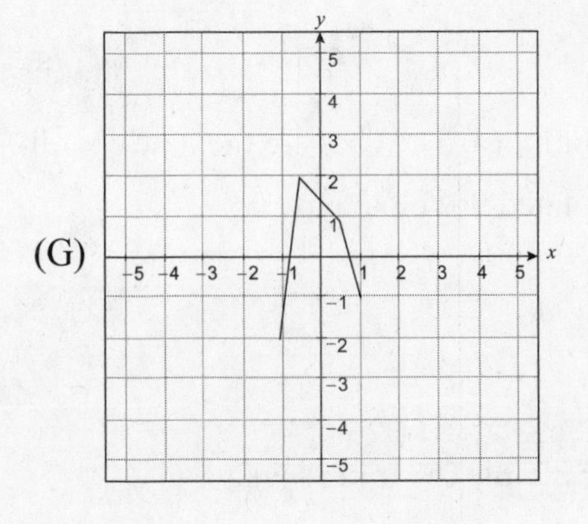

g(x) is horizontally compressed by a factor of 2.

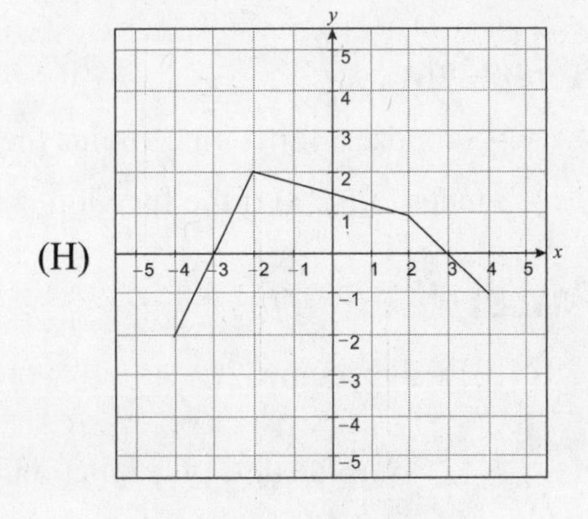

g(x) is horizontally elongated by a factor of 2.

A 90° **rotation** of a function moves each point P to a new point P' so that $OP = OP'$ and $\overline{OP}$ is perpendicular to $\overline{OP'}$.

The letter O represents the origin, which is located at (0, 0). If the rotation is *counterclockwise*, each point (x, y) becomes $(-y, x)$. If the rotation is *clockwise*, each point (x, y) becomes $(y, -x)$.

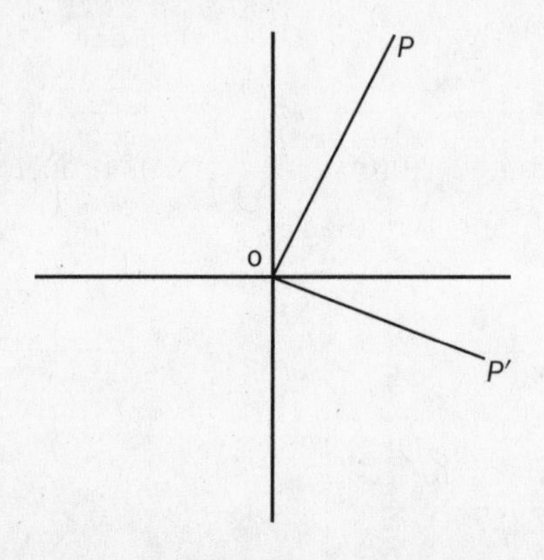

PROBLEM

Suppose a function contains the point (4, 3). What are the new coordinates of this point if the function is rotated 90° counterclockwise?

SOLUTION

(4, 3) will become (−3, 4).

PROBLEM

Suppose a function contains the point (4, 3). What are the new coordinates of (4, 3) if the function is rotated 90° clockwise?

SOLUTION

(4, 3) will become (3, −4).

A **reflection** of a function is simply the mirror image of the function.

A reflection about the x-axis changes point (x, y) into point $(x, -y)$. A reflection about the y-axis changes point (x, y) into point $(-x, y)$. A reflection about the line $y = x$ will move each point P to a new point P' so that the line $y = x$ is the perpendicular bisector of $\overline{PP'}$. Each point (x, y) becomes (y, x) after the reflection.

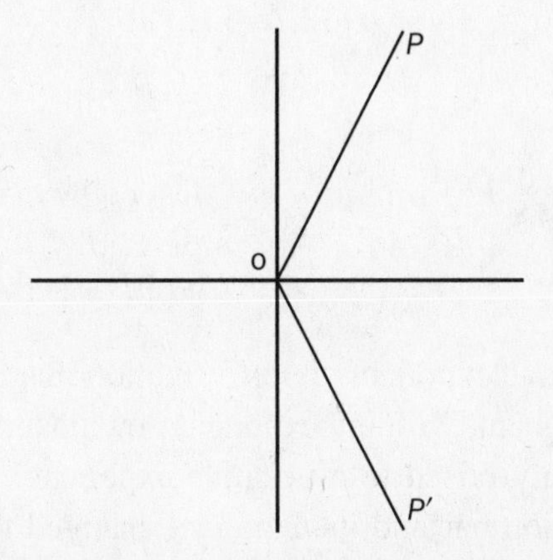

PROBLEM

A function contains the point $(-6, 3)$. If this function is reflected about the line $y = x$, what will be the new coordinates for $(-6, 3)$?

SOLUTION

$(-6, 3)$ will become $(3, -6)$.

EXPONENTS

When a number is multiplied by itself a specific number of times, it is said to be **raised to a power**. The way this is written is $a^n = b$, where a is the number or base; n is the **exponent** or **power** that indicates the number of times the base appears when multiplied by itself; and b is the **product** of this multiplication.

In the expression 3^2, 3 is the base and 2 is the exponent. This means that 3 appears 2 times when multiplied by itself (3×3), and the product is 9.

An exponent can be either positive or negative. A negative exponent implies a fraction, such that if n is a negative integer

$$a^{-n} = \frac{1}{a^n} a \neq 0$$

So, $2^{-4} = \frac{1}{2^4} = \frac{1}{16}$

The **reciprocal** of a number is 1 divided by that number. The exception is 0, since $\frac{1}{0}$ is undefined.

We see that a negative exponent yields a fraction that is the reciprocal of the original base and exponent, with the exponent now positive instead of negative. Essentially, any quantity raised to a negative exponent can be "flipped" to the other side of the fraction bar and its exponent changed to a positive exponent. For example,

$$4^{-3} = \frac{1}{4^3} = \frac{1}{64}$$

An exponent that is 0 gives a result of 1, assuming that the base itself is not equal to 0.

$$a^0 = 1, a \neq 0$$

An exponent can also be a fraction. If m and n are positive integers,

$$a^{\frac{m}{n}} = \sqrt[n]{a^m}$$

The numerator remains the exponent of a, but the denominator tells what root to take. For example,

$$4^{\frac{3}{2}} = \sqrt[2]{4^3} = \sqrt{64} = 8$$

$$3^{\frac{4}{2}} = \sqrt[2]{3^4} = \sqrt{81} = 9$$

If a fractional exponent is negative, the operation involves the reciprocal as well as the roots. For example,

$$27^{-\frac{3}{2}} = \frac{1}{27^{\frac{3}{2}}} = \frac{1}{\sqrt[3]{27^2}} = \frac{1}{\sqrt[3]{729}} = \frac{1}{9}$$

PROBLEMS

Simplify the following expressions:

1. -3^{-2}
2. $(-3)^{-2}$
3. $\frac{-3}{4^{-1}}$
4. $-16^{-\frac{1}{2}}$

SOLUTIONS

1. Here the exponent applies only to 3. Since

$$x^{-y} = \frac{1}{x^y}, \text{ so } -3^{-2} = -(3)^{-2} = -\left(\frac{1}{3^2}\right) = -\frac{1}{9}$$

2. In this case, the exponent applies to the negative base. Thus,

$$-(3)^{-2} = \frac{1}{(-3)^2} = \frac{1}{(-3)(-3)} = \frac{1}{9}$$

3. $$\frac{-3}{4^{-1}} = \frac{-3}{\left(\frac{1}{4}\right)^1} = \frac{-3}{\frac{1^1}{4^1}} = \frac{-3}{\frac{1}{4}} = \frac{-3}{1} \times \frac{4}{1} = -12$$

4. $$-16^{-\frac{1}{2}} = \frac{1}{16^{\frac{1}{2}}} = -\frac{1}{\sqrt{16}} = -\frac{1}{4}$$

GENERAL LAWS OF EXPONENTS

$a^p a^q = a^{p+q}$, bases must be the same

$$4^2 4^3 = 4^{2+3} = 4^5 = 1{,}024$$

$(a^p)^q = a^{pq}$

$$(2^3)^2 = 2^6 = 64$$

$\frac{a^p}{a^q} = a^{p-q}$, bases must be the same, $a \neq 0$

$$\frac{3^6}{3^2} = 3^{6-2} = 3^4 = 81$$

$(ab)^p = a^p b^p$

$$(3 \times 2)^2 = 3^2 \times 2^2 = (9)(4) = 36$$

$$\left(\frac{a}{b}\right)^p = \frac{a^p}{b^p}, b \neq 0$$

$$\left(\frac{4}{5}\right)^2 = \frac{4^2}{5^2} = \frac{16}{25}$$

EXPONENTIAL EQUATIONS

Most of the CLEP College Mathematics exam is about linear and polynomial functions, as discussed earlier in this chapter. However, **exponential functions**, in the form of $y=b^x$, are also tested on the exam.

The constant b is called the **base**, where b is a positive number. An exponential graph tends to increase or decrease rapidly because x is in the exponent. The following graphs are examples of exponential curves.

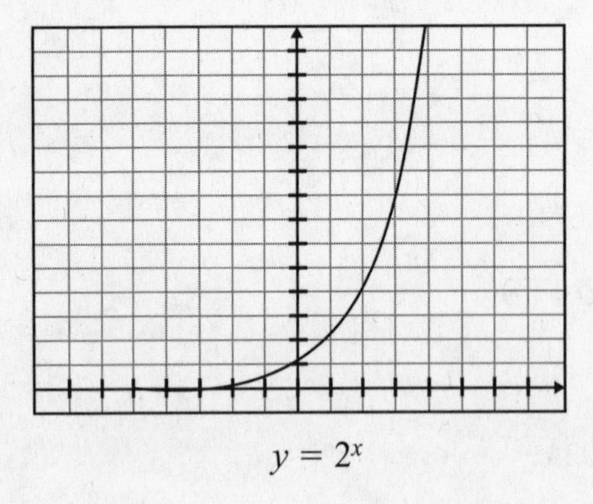

$y = 2^x$

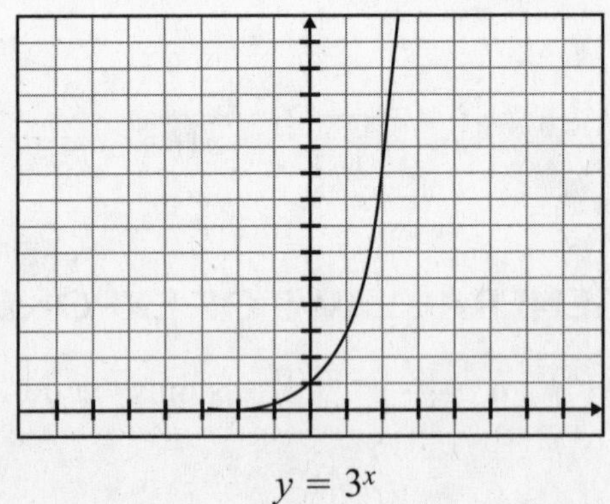

$y = 3^x$

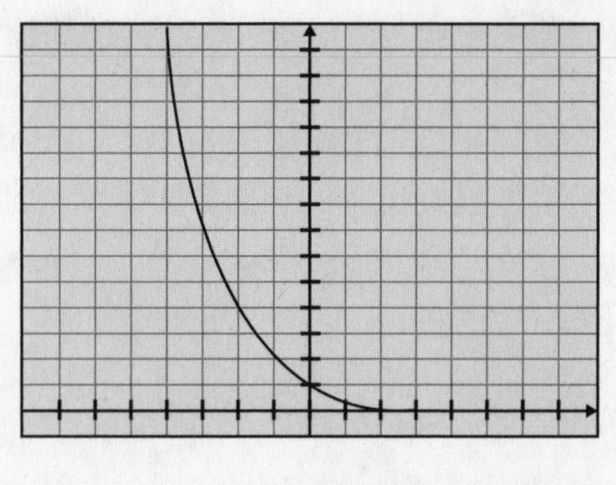

$y = 0.5^x$

A **growth curve** occurs when b > 1, and the larger b is, the steeper the growth curve is.

A **decay curve** occurs if $0 < b < 1$, as shown in the last graph above.

All exponential curves in the form of $y = b^x$, whether growth or decay, have certain features. The domain of the function is $(-\infty, \infty)$ since the exponent can be any number. However, the range of exponential functions is $(0, \infty)$ as any positive number raised to a positive power is positive and a positive number raised to a negative power creates a positive fraction. All exponential functions in the form of $y = b^x$ pass through the point (0, 1) as any base b raised to the zero power is one.

Solving basic exponential equations can be accomplished by using the fact that if $b^x = b^y$, then $x = y$. If the bases are the same, the powers must also be equal.

PROBLEM

Solve for x: $2^{x+1} = 8$

SOLUTION

We need to write 8 as a power of 2 so the bases are the same. We know $8 = 2^3$ so we have $2^{x+1} = 2^3$.

Thus $x + 1 = 3$, and $x = 2$.

PROBLEM

Solve for x: $3^{2x-3} = 9$

SOLUTION

$$3^{2x-3} = 3^2$$
$$2x - 3 = 2$$
$$2x = 5$$
$$x = \frac{5}{2}$$

PROBLEM

Solve for x: $3^{4x-1} = \frac{1}{3}$

SOLUTION

$$3^{4x-1} = 3^{-1}$$
$$4x - 1 = -1$$
$$x = 0$$

By using the facts for operations with exponents earlier in the chapter, we can solve more complicated exponential equations.

PROBLEM

Solve for x: $4^{3x-3} = 8^{x+2}$

SOLUTION

We cannot write 8 as a power of 4 but we can write both 4 and 8 as powers of 2.

$$\left(2^2\right)^{3x-3} = \left(2^3\right)^{x+2}$$

(Remember that powers raised to powers means to multiply the powers.)

$$2^{6x-6} = 2^{3x+6}$$
$$6x - 6 = 3x + 6$$
$$3x = 12$$
$$x = 4$$

LOGARITHMS

Solving exponential equations like the ones above are fairly easy when each side of the equation has common bases. But that technique doesn't work with $3^{x-1} = 6$ as we cannot write 6 as a power of 3.

To solve problems of this type, we introduce **logarithms** (also referred to as **logs**). A logarithm is the inverse of an exponential function. We know that when we find an inverse, we interchange x and y. So if $y = b^x$, the inverse is $x = b^y$. The use of logarithms helps us to solve exponential functions for y, since the notation $\log_b y = x$ is the same as $b^x = y$. The base b cannot be negative.

What you need to know about logarithms for the CLEP exam is this:

$\log_b y = x$ is the same thing as saying $b^x = y$.

When you are asked to find a logarithm (the abbreviation of *logarithm* is *log*), call it x. Then write the expression as an exponential and solve it using the technique above.

PROBLEM

Find $\log_2 8$.

SOLUTION

We do not know $\log_2 8$, so we write $\log_2 8 = x$.

Now we can write write the statement exponentially as $2^x = 8$.

Then write the right side as a power of 2: $2^x = 2^3$

Then $x = 3$. So, $\log_2 8 = 3$.

PROBLEM

Find $\log_3 81$.

SOLUTION

$\log_3 81 = x$

$3^x = 81$

$3^x = 3^4$

$x = 4$

PROBLEM

Find $\log_4 32$.

SOLUTION

$$\log_4 32 = x$$
$$4^x = 32$$
$$2^{2x} = 2^5$$
$$2x = 5$$
$$x = \frac{5}{2}$$

PROBLEM

Find $\log_8 \sqrt{2}$.

SOLUTION

$$\log_8 \sqrt{2} = x$$
$$8^x = \sqrt{2}$$
$$2^{3x} = 2^{\frac{1}{2}}$$
$$3x = \frac{1}{2}$$
$$x = \frac{1}{2}\left(\frac{1}{3}\right) = \frac{1}{6}$$

PROBLEM

Find $\log_9 1$.

SOLUTION

$$\log_9 1 = x$$
$$9^x = 1$$
$$x = 0$$

Common Logarithms

If the base is not specified, it is assumed to be 10. A logarithm with a base of 10 is called a **common logarithm**. The notations $\log_{10} x$ and $\log x$ are the same.

PROBLEM

Find $\log 1000$.

SOLUTION

$\log 1000 = x$

Since this is the same as $\log^{10} 1000 = x$,

$10^x = 1000$

$10^x = 10^3$

$x = 3$, or $\log 1000 = 3$

PROBLEM

Find $\log 1$.

SOLUTION

$\log 1 = x$

$10^x = 1$

$10^x = 10^0$

$x = 0$

Logarithm Rules

There are three basic rules for operations with logarithms that are important. These rules work with logs to any base. They are as follows:

Rule 1. $\log a + \log b = \log(a \cdot b)$

Rule 2. $\log a - \log b = \log\left(\frac{a}{b}\right)$

Rule 3. $\log a^b = b \log a$

PROBLEM

Find the value of $\log 25 + \log 4$.

SOLUTION

Even though we cannot find the log of 25 or the log of 4, we can use rule 1 to say

$\log 25 + \log 4 = \log(25 \cdot 4) = \log 100$

$\log 100 = x$

$10^x = 100$

$10^x = 10^2$

$x = 2$, or $\log 25 + \log 4 = 2$

PROBLEM

Find the value of $\log_2 80 - \log_2 5$.

SOLUTION

Even though we cannot find the $\log_2 80$ or $\log_2 5$, we can use rule 2 to say

$\log_2 80 - \log_2 5 = \log_2\left(\frac{80}{5}\right) = \log_2 16$

$\log_2 16 = x$

$2^x = 16$

$2^x = 2^4$

$x = 4$, or $\log_2 80 - \log_2 5 = 4$

PROBLEM

Find $\log 10^{35}$.

SOLUTION

We can use rule 3 to say $\log 10^{35} = 35 \log 10$

$\log 10 = x$

$10^x = 10$

$x = 1$

Since $35(1) = 35$, $\log 10^{35} = 35$.

LINEAR AND EXPONENTIAL GROWTH

LINEAR GROWTH

Linear equations and functions are used to model situations in which a quantity y grows or declines the same amount over the same time step or change in x. The graph of a linear growth model will be a straight line. We will usually be given the slope m and the y-intercept b and the linear function is given by $f(x)=mx+b$.

PROBLEM

Suppose that a car uses 0.03 gallons of gas per mile and the fuel tank, which holds 15 gallons of gas, is full. Using this information, we can determine a linear model for the remaining amount of fuel in the gas tank after driving x miles.

SOLUTION

Recall that a linear function is one that can be written in the form $f(x)=mx+b$, where m is the slope of the line and b is the y-intercept. The slope is the rate at which the car is using gas, 0.03 gallons per mile. Because the car is using the fuel, the amount of fuel in the tank is decreasing. Therefore, the slope is negative, and we have $m = -.03$.

When $x = 0$ (before the car drives away from the pump), there are 15 gallons of gas in the tank. The y-intercept is (0, 15).

So our function in the form of $f(x)=mx+b$ can be written as $f(x)=-0.03x+15$. We can graph this function as shown below.

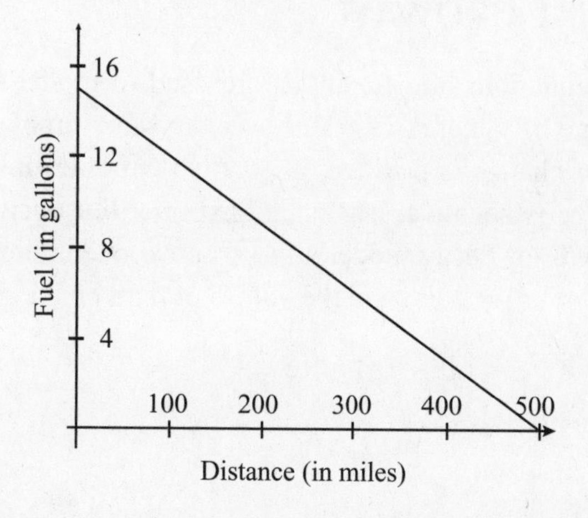

To find how far the car can travel on a tank of gas, we set the function $f(x) = -0.03x + 15$ equal to zero because when there are 0 gallons of fuel, the car cannot go any further.

$$f(x) = -0.03x + 15 = 0$$

$$15 = 0.03x$$

$$x = \frac{15}{0.03} = 500 \text{ miles}$$

The car can go 500 miles on a full tank of gas.

A problem like this is a **linear decay** problem.

PROBLEM

A cellular phone company offers several different service options. One option, for people who plan on using the phone only in emergencies, costs the user \$8.95 a month plus \$0.33/minute for each minute the phone is used. Write a linear growth function for the monthly cost of the phone in terms of the number of minutes the phone is used. If the phone is used a total of 25 minutes, what would be the monthly cost?

SOLUTION

If x represents the number of minutes used, the cost is $8.95 + 0.33x$.

$$f(x) = 0.33x + 8.95$$

$$f(25) = 0.33(25) + 8.95 = 17.20$$

It will cost \$17.20 for 25 minutes of cell phone use.

EXPONENTIAL GROWTH

Exponential equations and functions are used to model situations in which the rate of change of y increases faster over the same time step or change in x. When the rate of change is positive, it is called **exponential growth** and when the rate of change is negative, it is called **exponential decay**. The graph of an exponential growth or decay model will be an exponential curve. The equation is in the form of $f(x) = a \cdot b^x$ and the values of a and b are generally given.

Example:

A diamond merchant has determined the values of several white diamonds that have different weights, measured in carats, but are similar in quality. The value of these diamonds is given in the chart below.

Weight (carats)	0.50	0.75	1.00	1.25	1.50	1.75	2.00	3.00	4.00
Value	\$4,700	\$5,200	\$5,800	\$6,300	\$7,000	\$7,700	\$8,500	\$12,400	\$18,300

The exponential growth function that models the value of the diamonds as a function of their weights is $\text{Value} = 3912(1.47)^{\text{carat weight}}$. We can use this model to approximate the value of a 2.5-carat diamond: $\text{Value} = 3912(1.47)^{2.5} = \$10,249$. The Millenium Star is one of the world's largest diamonds at 203 carats. We should not expect to use the $\text{Value} = 3912(1.47)^{\text{carat weight}}$ equation because the Millenium Star is a different type of diamond and a much greater weight than the equation that modeled it. But if it were applied, the cost of this diamond would be $\text{Value} = 3912(1.47)^{203} = \$3.6 \times 10^{37}$. That is why these rare gems are priceless.

PROBLEM

The table below shows the decreasing cost of hard-disk storage space.

Year	1992	1997	2002	2007	2012
Cost per megabyte	\$915	\$50.39	\$2.78	\$0.15	\$0.01

The exponential decay function is given by Cost per megabyte = $915 \times 0.56^{\text{year} - 1992}$. What was the approximate cost per megabyte of disk storage in the year 2005? What will be the approximate cost per megabyte of disk storage in the year 2020?

SOLUTION

In the year 2005, the

$$\text{Cost per megabyte} = 915 \times 0.56^{2005-1992} = 915 \times 0.56^{13} = 0.487 \approx \$0.49.$$

It isn't advisable to use this formula for the year 2020 as no one can forecast whether the cost of hard disk space will adhere to this formula in the future. But the value is given by:

$$\text{Cost per megabyte} = 915 \times 0.56^{2020-1992} = 915 \times 0.56^{28} = 8.14 \times 10^{-5} = \$0.0000814.$$

Drill Questions

1. Which one of the following is an equation of a line parallel to the graph of $5x + 9y = 14$?

(A) $y = \frac{5}{9}x + 14$

(B) $y = -\frac{5}{9}x - 14$

(C) $y = -\frac{9}{5}x + 14$

(D) $y = -\frac{9}{5}x - 14$

2. The solution set of the inequality $5 - 7x \geq -9$ is

(A) $X = \{x \mid x \leq 2\}$

(B) $X = \{x \mid x \geq 2\}$

(C) $X = \left\{x|x \leq \frac{4}{7}\right\}$

(D) $X = \left\{x|x \geq \frac{4}{7}\right\}$

3. Place the following in order from largest to smallest.

I. $\log 5000 - \log 5$ II. $\log 50 + \log 2$ III. $\log 10^4$

(A) I, II, III

(B) II, I, III

(C) III, I, II

(D) I, III, II

4. The graph below is described by what inequity?

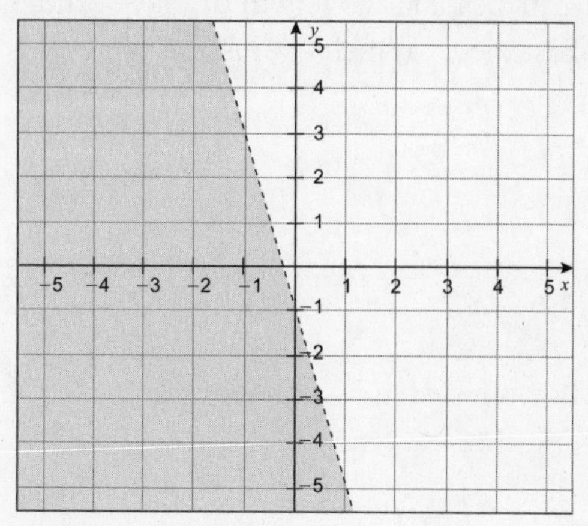

(A) $y > 1 - 4x$
(B) $y \geq 1 - 4x$
(C) $y < -4x - 1$
(D) $y \leq -4x - 1$

5. What is the solution set for x in the following inequality?

$$-6 < \frac{2}{3}x + 6 < 2$$

(A) $-12 < x < 0$
(B) $-6 < x < 0$
(C) $-18 < x < -12$
(D) $-18 < x < -6$

6. What is the simplified expression for $-5i^7 + (2i)(i^2 - 2)$?

(A) $11i$
(B) $-11i$
(C) i
(D) $-i$

7. If $x + 2$ is a factor of $3x^3 - x + k$, what is the value of k?

(A) 22
(B) 16
(C) -16
(D) -22

8. A piece of lumber that is 28 inches in length is divided into three pieces. The length of the first piece is x inches and the length of the second piece is four times the length of the first piece. Which expression represents the length of the third piece?

(A) $28 - 4x$
(B) $28 - 5x$
(C) $4x - 28$
(D) $5x - 28$

9. What is the value of y in the following system of equations?

$4x + 2y = 5$

$3x + 5y = 9$

(A) $\frac{1}{2}$
(B) $\frac{3}{2}$
(C) $\frac{5}{2}$
(D) $\frac{7}{2}$

10. $2^5 = 32$ is equivalent to which logarithmic expression?

(A) $\log_5 32 = 2$
(B) $\log_{32} 5 = 2$
(C) $\log_2 32 = 5$
(D) $\log_2 5 = 32$

11. If $f(x) = 2x + 3$ and $g(x) = x^3 - 5$, what is the value of $f(g(-3))$?

(A) -61
(B) -32
(C) 25
(D) 47

12. If $h(x) = \frac{5}{x-2}$, which of the following is equivalent to the inverse of $h(x)$?

(A) $\frac{x-2}{5}$
(B) $\frac{5}{x+2}$
(C) $\frac{2x-5}{5}$
(D) $\frac{2x+5}{5}$

13. What is the domain of $f(x) = \sqrt{6-3x}$?

(A) $x \geq 2$
(B) $x \leq 2$
(C) $x \geq -2$
(D) $x \leq -2$

14. If $\{(4, -1), (5, -3), (6, -9), (\_\_, -16)\}$ represents a function, which one of the following *cannot* be used to fill in the blank space?

(A) 5
(B) 3
(C) -4
(D) -9

15. Each point of the function $f(x) = x^2 + 10$ is moved 3 units to the left and 2 units down to create a new function $g(x)$. What would be the y-coordinate of a point on the graph of $g(x)$ whose x-coordinate is -1 on the graph of $f(x)$?

(A) 6
(B) 7
(C) 8
(D) 9

16. If $f(x)$ is a linear function such that $f(1) = 5$ and $f(4) = -7$, then what is the value of $f(-1)$?

(A) 11
(B) 12
(C) 13
(D) 14

17. The function $g(x)$ is known to have a range of all real numbers except zero. Which one of the following expressions could represent $g(x)$?

(A) $4x^2 - 4$

(B) $\frac{4}{x}$

(C) $\frac{4}{x^2}$

(D) $-\sqrt{x^2 - 4}$

18. A builder estimates that the cost to build a new home is $50,000 plus $90 per square foot of floor space in the house. Use a linear model to determine the size of a house that can be built for $500,000.

(A) 500
(B) 2,000
(C) 5,000
(D) 6,111

19. The point $(-2, -3)$ is rotated 90° clockwise about the origin. What is its new location?

(A) $(-3, 2)$
(B) $(-3, -2)$
(C) $(3, -2)$
(D) $(3, 2)$

20. For which one of the following functions is $f(1) = f(-1) = 2$?

(A) $f(x) = x^2 - 3x + 4$
(B) $f(x) = 4x^2 - 2$
(C) $f(x) = x^2 - 2x - 1$
(D) $f(x) = 3x^2 - x$

Answers to Drill Questions

1. **(B)** Rewrite the equation $5x + 9y = 14$ in slope-intercept form. Then $9y = -5x + 14$, which leads to $y = -\frac{5}{9}x + \frac{14}{9}$, where $-\frac{5}{9}$ is the slope and $\frac{14}{9}$ is the y-intercept. The equation of any line parallel to the graph of $5x + 9y = 14$ must also have a slope of $-\frac{5}{9}$. Only Choice (B) satisfies this condition.

2. **(A)** Subtract 5 from each side of $5 - 7x \geq -9$, so that $5 - 7x - 5 \geq -9 - 5$, which simplifies to $-7x \geq -14$. Next, divide each side by -7. Since we are dividing by a negative number, we must switch the order of the inequality. Then $\frac{-7x}{-7} \leq \frac{-14}{-7}$, which becomes $x \leq 2$. This is equivalent to $X = \{x \mid x \leq 2\}$.

3. **(C)** I. $\log 5000 - \log 5 = \log \frac{5,000}{5} = \log 1,000$: $10^x = 1,000 \Rightarrow x = 3$
   II. $\log 50 + \log 2 = \log (50 \times 2) = \log 100$: $10^x = 100 \Rightarrow x = 2$
   III. $\log 10^4 = 4 \log 10 = 4(1) = 4$

4. **(C)** The slope is -4 and the y- intercept is -1. Since the shading is below the line and the line is dashed, the inequality is $y < -4x - 1$.

5. **(D)** Subtract 6 from each part of the double inequality to get $-6 - 6 < \frac{2}{3}x + 6 - 6 < 2 - 6$, which simplifies to $-12 < \frac{2}{3}x < -4$. Now divide each expression by $\frac{2}{3}$, which is equivalent to multiplying by $\frac{3}{2}$. Thus, $(-12)\left(\frac{3}{2}\right) < \left(\frac{2}{3}x\right)\left(\frac{3}{2}\right) < (-4)\left(\frac{3}{2}\right)$, which leads to $-18 < x < -6$.

6. **(D)** Using the Distributive Law, $-5i^7 + (2i)(i^2 - 2) = -5i^7 + 2i^3 - 4i$. Recall that the powers of i are cyclical in groups of 4, which means that $i = i^5 = i^9 = \ldots, -1 = i^2 = i^6 = i^{10} = \ldots, -i = i^3 = i^7 = i^{11} = \ldots$, and $1 = i^4 = i^8 = i^{12} = \ldots$ Thus, $-5i^7 + 2i^3 - 4i = (-5)(-i) + (2)(-i) - 4i = 5i - 2i - 4i = -i$.

7. **(A)** Based on the Factor Theorem, if $x + 2$ is a factor of $3x^3 - x + k$, then -2 must be a solution for x in the equation $3x^3 - x + k = 0$. This means that $(3)(-2)^3 - (-2) + k = 0$. This equation simplifies to $(3)(-8) + 2 + k = 0$. Thus, $k = 24 - 2 = 22$.

8. **(B)** The length of the first piece is x and the length of the second piece is $4x$. The length of the third piece is found by adding the first two pieces and subtracting this sum from the total of 28 inches. The sum of the first two pieces is $5x$, so the length of the third piece must be $28 - 5x$.

9. **(B)** We can eliminate the variable x from the given system of equations as follows: Multiply the first equation by 3 to get $12x + 6y = 15$. Next, multiply the second equation by 4 to get $12x + 20y = 36$. Now, by the subtraction property of equalities, $(12x + 6y) - (12x + 20y) = 15 - 36$. This equation simplifies to $-14y = -21$. Thus, $y = \frac{-21}{-14} = \frac{3}{2}$.

10. **(C)** By definition, the expression $\log_b x = y$ is equivalent to $b^y = x$. Substituting 2 for b, 32 for x, and 5 for y, we conclude that $2^5 = 32$ is equivalent to $\log_2 32 = 5$.

11. **(A)** $g(-3) = (-3)^3 - 5 = -27 - 5 = -32$. Thus, $f(g(-3)) = f(-32) = (2)(-32) + 3 = -64 + 3 = -61$.

12. **(D)** The inverse of a function is found by reversing the roles of the two variables, then solving for the new dependent variable. Let $y = h(x)$ so that the initial function can be written as $x = \frac{5}{y-2}$. Interchanging the x and y, we get $x = \frac{5}{x-2}$. Multiply this equation by $y - 2$ to get $(x)(y - 2) = 5$. This equation simplifies to $xy - 2x = 5$. Next, add $2x$ to each side to get $xy - 2x + 2x = 5 + 2x$, which becomes $xy = 5 + 2x$. Finally, divide both sides by x to get $y = \frac{5+2x}{x}$. The right side of this equation, which is equivalent to $\frac{2x+5^x}{x}$, represents the expression for the inverse of $h(x)$.

13. **(B)** The domain of $f(x) = \sqrt{6-3x}$ is defined as the allowable values of x. The square root of any function must be at least zero in order to represent a real value. The domain is found by solving $6 - 3x \geq 0$. Subtracting 6 from

each side leads to $-3x \geq -6$. Finally, divide both sides by – 3 and reverse the order of the inequality. Thus, $\frac{-3x}{-3} \leq \frac{-6}{-3}$, which becomes $x \leq 2$.

14. **(A)** The definition of a function, as it relates to a set of ordered pairs, is that any specific first number must be associated with a single second number. This implies that a set of ordered pairs is not a function if two ordered pairs contain the same first number but a different second number. The given set contains the elements (4, −1), (5, −3), (6, −9), and (_, −16). In order for this set to qualify as a function, we cannot repeat 4, 5, or 6 as a first number of the last ordered pair.

15. **(D)** Substitute −1 for x in the function f(x) to get $(-1)^2 + 10 = 1 + 10 = 11$. The corresponding point on the graph of $f(x)$ is (−1, 11). In order to find the corresponding point for $g(x)$, this point will be moved 3 units to the left and 2 units down. Thus, the point (−1, 11) will become the point (−4, 9) on the graph of $g(x)$. So, the y-coordinate becomes 9.

16. **(C)** The points on the graph of this linear function are (1, 5) and (4, −7). The function can be written as $y = mx + b$, where m is the slope and b is the y-intercept. By substitution of the two given points, we get $5 = (m)(1) + b$ and $-7 = (m)(4) + b$. Now subtracting the second of these equations from the first equation leads to $5 - (-7) = (m)(1) + b - (m)(4) - b$, which becomes $12 = -3m$. So $m = -4$. Returning to the equation $5 = (m)(1) + b$, we can substitute – 4 for m so that $5 = (-4)(1) + b$. Then $b = 9$. The equation of this linear function is $y = -4x + 9$. Let's replace y with $f(x)$. Finally, $f(-1) = (-4)(-1) + 9 = 4 + 9 = 13$.

17. **(B)** The range is represented by the $g(x)$ values. For $g(x) = \frac{4}{x}$, $g(x)$ can assume any value (including negative numbers) except zero. Also, note that $x \neq 0$. A graph of $g(x) = \frac{4}{x}$ would also confirm that the range is all numbers except zero.

18. **(C)** Price = 50,000 + 90(square feet) = 500,000
    90(square feet) = 450,000
    square feet $= \frac{450{,}000}{90} = 5{,}000$

19. **(A)** If the point (x, y) is rotated 90° clockwise about the origin, its new location is given by $(y, -x)$. Let $x = -2$ and $y = -3$. Then for the point

$(-2, -3)$, its new location after being rotated 90° clockwise about the origin is found by interchanging the coordinates, then switching the sign of the new second coordinate. The correct answer is $(-3, 2)$.

20. **(B)** The quickest way to solve this problem is by substitution into each answer choice. For choice (A), $f(1) = 1^2 - (3)(1) + 4 = 2$, but $f(-1) = (-1)^2 - (3)(-1) + 4 = 8$. So, choice (A) is incorrect. Choice (B) is correct because $f(1) = 4(1)^2 - 2 = 2$ and $f(-1) = 4(-1)^2 - 2 = 2$. Note that for choice (C), $f(1) = -2$, but $f(-1) = 2$. For choice (D), $f(1) = 2$, but $f(-1) = 4$.

CHAPTER 4
Counting and Probability

CHAPTER 4

COUNTING AND PROBABILITY

SAMPLE SPACES AND COUNTING

In an event or activity that has a few or many possible outcomes, a sample space lists all of those possible outcomes.

Example:

A coin is tossed. The sample space is {heads, tails}.

Example:

A woman is expecting a delivery of a sofa between 1pm and 2pm. In terms of delivery time, the sample space is {early, on time, late, no show}.

PROBLEM

A die is rolled. Write the sample space.

SOLUTION

$$\{1, 2, 3, 4, 5, 6\}$$

PROBLEM

A piggy bank contains many nickels, dimes, and quarters. Two coins are chosen at random. Write the sample space of possible sums.

SOLUTION

$$\{10, 15, 30, 20, 35, 50\}$$

THE COUNTING PRINCIPLE

When dealing with the occurrence of more than one event or activity, it is important to be able to quickly determine how many outcomes exist, rather than listing them all in the sample space. If there are "m ways" for one activity to occur and "n ways" for a second activity to occur, the counting principle states that there are "$m \cdot n$ ways" for both to occur.

Example:

A vending machine sells snacks and soft drinks. If there are 8 different snacks and 5 different soft drinks, there are $8 \cdot 5 = 40$ different ways to choose a snack and a soft drink.

Example:

A popular car comes in 10 different colors, 4 different interiors, 3 engine sizes, and with or without a navigation system. There are $10 \cdot 4 \cdot 3 \cdot 2 = 240$ configurations of the car that are possible.

PROBLEM

The menu for a ship's dining room has 6 different appetizers, 6 different entrées, and 4 different desserts. How many different meals can a passenger order?

SOLUTION

$$6 \cdot 6 \cdot 4 = 144 \text{ meals}$$

PROBLEM

When a new area code is introduced, how many 7-digit telephone numbers are available, assuming that the first digit cannot start with a 0 or a 1?

SOLUTION

$$8(10)(10)(10)(10)(10)(10) = 8{,}000{,}000 \text{ telephone numbers}$$

FACTORIAL

The factorial of a positive integer n, written as $n!$, is the product of all positive integers less than or equal to n.

$5! = 5 \cdot 4 \cdot 3 \cdot 2 \cdot 1 = 120$

$8! = 8 \cdot 7 \cdot 6 \cdot 5 \cdot 4 \cdot 3 \cdot 2 \cdot 1 = 40{,}320$

$$\frac{6!}{3!} = \frac{6 \cdot 5 \cdot 4 \cdot 3 \cdot 2 \cdot 1}{3 \cdot 2 \cdot 1} = 120$$

0! is defined as 1.

PROBLEMS

Find the following:

(A) $6!$ (B) $\frac{7!}{3!}$ (C) $\frac{10!}{9!}$ (D) $\frac{10!}{5!\,5!}$

SOLUTIONS

(A) $6! = 6(5)(4)(3)(2)(1) = 720$

(B) $$\frac{7!}{3!} = \frac{7\ 6\ 5\ 4\ 3\ 2\ 1}{3\ 2\ 1} = 840$$

(C) $$\frac{10!}{9!} = \frac{10(9)(8)(7)(6)(5)(4)(3)(2)(1)}{9(8)(7)(6)(5)(4)(3)(2)(1)} = 10$$

(D) $$\frac{10!}{5!5!} = \frac{10\ 9\ 8\ 7\ 6\ 5\ 4\ 3\ 2\ 1}{5\ 4\ 3\ 2\ 1\ 5\ 4\ 3\ 2\ 1} = 252$$

PERMUTATIONS

A **permutation** is a set of objects in which position (or order) is important. An example is a "combination" to a safe, which is really a permutation. If the "combination" is 5-7-8, the order is important. 8-5-7 will not open the safe: it has to be exactly 5-7-8.

Permutations with repetition: An object *can* be repeated. In the lock example, it is possible for a "combination" to be 4-4-2 or 9-9-9. If we have n

things to choose from each time, the counting principle says that there will be $n \cdot n \cdot n \ldots$ permutations.

Example:

In the lock example, there are 10 possibilities (0, 1, … 9) and we want 3 of them. There are $10 \cdot 10 \cdot 10 = 10^3 = 1{,}000$ permutations. Note that this assumes that it is possible for a number to repeat.

Permutations without repetition: An object cannot be repeated. Once we choose it, it can no longer be chosen.

Example:

In the lock example above, suppose that you cannot use a number over again. Once you use it, you cannot re-use it. The first number has a choice of 10. Once you use it, you now have only 9 choices for the second number. Once you use it, you only have 8 choices for the third number. So there are $10 \cdot 9 \cdot 8 = 720$ permutations.

PROBLEM

An ice cream shop has 25 flavors. If a person orders a triple-decker cone, how many different cones are possible if (A) it is allowed to repeat a flavor and (B) no repetition of flavors is allowed?

SOLUTIONS

(A) $25(25)(25) = 15{,}625$

(B) $25(24)(23) = 13{,}800$

PROBLEM

A computer password has 5 characters. Digits and letters (both lowercase and uppercase) can be used. How many different passwords are available if (A) it is allowed to repeat a character and (B) no repetition of characters is allowed? Do not actually compute the answer.

SOLUTIONS

(A) 62(62)(62)(62)(62)

(B) 62(61)(60)(59)(58)

COMBINATIONS

A combination is a set of objects in which position (or order) is *not* important. An example would be a fruit salad that is a combination of cantaloupe, honeydew, and watermelon. We don't care what order the fruits are in. Watermelon, cantaloupe, and honeydew is the same fruit salad as honeydew, watermelon, and cantaloupe.

The best way to determine the number of combinations possible is to use a formula. When you have "n" objects and we wish to choose "r" of them, there are ${}_nC_r = \dfrac{n!}{r!(n-r)!}$ combinations possible. An alternate notation for combinations is $\dbinom{n}{r}$. When calculating combinations using this formula, we can usually do cancellation to make the calculations easier.

Example:

A basketball team has 9 members and 5 players are on the floor. The number of combinations of team members on the floor that are possible is given by:

$${}_9C_5 = \frac{9!}{5!(9-5)!} = \frac{9!}{5!\cdot 4!} = \frac{9\cdot 8\cdot 7\cdot 6\cdot 5\cdot 4\cdot 3\cdot 2\cdot 1}{5\cdot 4\cdot 3\cdot 2\cdot 1\cdot 4\cdot 3\cdot 2\cdot 1} = \frac{9\cdot 8\cdot 7\cdot 6\cdot \cancel{5}\cdot \cancel{4}\cdot \cancel{3}\cdot \cancel{2}\cdot \cancel{1}}{\cancel{5}\cdot \cancel{4}\cdot \cancel{3}\cdot \cancel{2}\cdot \cancel{1}\cdot 4\cdot 3\cdot 2\cdot 1} = 126$$

Example:

7 boys are on a camping trip and have two tents. 4 boys go into one tent and 3 boys go into the other. The number of ways to divide the boys is given by:

$${}_7C_4 = \frac{7!}{4!(7-4)!} = \frac{7!}{4\cdot 3!} = \frac{7\cdot 6\cdot 5\cdot 4\cdot 3\cdot 2\cdot 1}{4\cdot 3\cdot 2\cdot 1\cdot 3\cdot 2\cdot 1} = \frac{7\cdot \cancel{6}\cdot 5\cdot \cancel{4}\cdot \cancel{3}\cdot \cancel{2}\cdot \cancel{1}}{\cancel{4}\cdot \cancel{3}\cdot \cancel{2}\cdot \cancel{1}\cdot \cancel{3}\cdot \cancel{2}\cdot 1} = 35$$

It is important to be able to determine whether a problem defines a permutation or a combination.

Permutation	Combination
Picking players to pitch, catch, and play shortstop from a group of players.	Picking three team members from a group of players.
In a dog show, choosing 1st place, 2nd place, and 3rd place from a group of dogs.	In a dog show, choosing 3 dogs that will go to the finals from a group of dogs.
From a color paint brochure, choosing a color for the walls and a color for the trim.	From a color paint brochure, choosing two colors to paint the room.

PROBLEM

An essay exam has 10 questions and students are instructed to answer exactly 4 of them. How many ways can this be done?

SOLUTION

$$_{10}C_4 \text{ or } \binom{10}{4} = \frac{10!}{4! \cdot 6!} = \frac{10 \cdot 9 \cdot 8 \cdot 7 \cdot 6 \cdot 5 \cdot 4 \cdot 3 \cdot 2 \cdot 1}{4 \cdot 3 \cdot 2 \cdot 1 \cdot 6 \cdot 5 \cdot 4 \cdot 3 \cdot 2 \cdot 1} = 210$$

PROBLEM

A law firm has 25 associate lawyers and 2 of them will be promoted to a partner in the firm. How many ways can this be done?

SOLUTION

$$_{25}C_2 \text{ or } \binom{25}{2} = \frac{25!}{2! \cdot 23!} = \frac{25 \cdot 24 \cdot 23 \cdot 22 \cdot \ldots \cdot 3 \cdot 2 \cdot 1}{2 \cdot 1 \cdot 23 \cdot 22 \cdot \ldots \cdot 3 \cdot 2 \cdot 1} = \frac{25 \cdot 24}{2 \cdot 1} = 300$$

PROBLEM

A golfer owns 17 golf clubs. When he plays a tournament he is allowed only 14 clubs in his golf bag. How many ways can he select the clubs if one of the clubs must be a putter?

SOLUTION

$${}_{16}C_{13} \text{ or } \genfrac{}{}{0pt}{}{16}{13} = \frac{16!}{13!\ 3!} = \frac{16\cdot15\cdot14\cdot13\cdot12\ \ldots\ 3\cdot2\cdot1}{13\cdot12\ \ldots\ 3\cdot2\cdot1\cdot3\cdot2\cdot1} \quad \frac{16\cdot15\cdot14}{3\cdot2\cdot1} \quad 560$$

PROBABILITY

Probability refers to how likely an event is to occur. The probability of an event is a number between 0 and 1 inclusive. A probability of 0 means the event cannot occur, a probability of 1 means the event must occur, and a probability of 0.5 means that the event is as likely to occur than not. Probability can be expressed as a fraction, decimal, or percent.

Example:

A coin having two heads is tossed. The probability of heads appearing is 1. The probability of tails appearing is 0.

Example:

A fair coin is tossed. The probability of heads occurring is 0.5. The probability of tails occurring is 0.5.

In general, the probability of an event happening

$$= \frac{\text{number of ways the event can happen}}{\text{total number of outcomes}}.$$

Example:

A class has 20 students with 14 boys and 6 girls. The teacher calls on a student at random. The probability that she chooses a boy is $\frac{14}{20} = \frac{7}{10} = 0.7 = 70\%$.

Example:

From a group of 15 whiteboard markers, 10 are dried up. If I choose a marker at random, the probability that I choose a marker that works is $\frac{5}{15}=\frac{1}{3}=0.\bar{3}=33.\bar{3}\%$.

Example:

5 people line up at random and one of them in named Jack. The probability that Jack is first in line is $\frac{1}{5}$.

A more difficult example using permutations

5 people line up at random. If Jack and Jill are boyfriend and girlfriend, they wish to find the probability that they will be next to each other

Probability that Jack and Jill are together

$$=\frac{\text{Number of ways Jack and Jill can be together}}{\text{Number of ways 5 people can line up}}$$

Number of ways 5 people can line up $= 5! = 5 \cdot 4 \cdot 3 \cdot 2 \cdot 1 = 120$

It is easy to simply generate all the possibilities with Jack and Jill together.

1	2	3	4	5
Jack	Jill			
	Jack	Jill		
		Jack	Jill	
			Jack	Jill

1	2	3	4	5
Jill	Jack			
	Jill	Jack		
		Jill	Jack	
			Jill	Jack

Probability that Jack and Jill are together $=\frac{8}{120}=\frac{1}{15}=0.0\bar{6}=6.\bar{6}\%$

A more difficult example with combinations

There are 5 ice cream flavors, two of which are chocolate and vanilla. I choose 3 different flavors at random. What is the probability that I have both chocolate and vanilla?

Probability of having chocolate and vanilla

$$= \frac{\text{Number of dishes with chocolate, vanilla, and 1 other flavor}}{\text{Number of dishes with 3 flavors}}$$

Number of dishes with chocolate, vanilla, and 1 other flavor = 3

Number of dishes with 3 flavors = ${}_5C_3 = \frac{5!}{3!\,2!} = \frac{5 \cdot 4 \cdot 3 \cdot 2 \cdot 1}{3 \cdot 2 \cdot 1 \cdot 2 \cdot 1} = 10$

Probability of having chocolate and vanilla = $\frac{3}{10} = 0.3 = 30\%$.

When there are relatively few outcomes possible, a sample space can be made and the probability of an event determined by counting the number of objects in the sample space as well as the number of objects in the sample space that have the characteristics you wish. For example, suppose a day is chosen at random. What is the probability that the day is on a weekend? The sample space is:

{Monday, Tuesday, Wednesday, Thursday, Friday, Saturday, and Sunday}. There are 7 objects in the sample space and 2 of them are on the weekend, so the probability is $\frac{2}{7}$.

Still, it can be easy to make a mistake. If we define the sample space as {weekday, weekend day}, you might think the probability of choosing a weekend day as $\frac{1}{2}$, which is incorrect. So if you use the sample space technique to determine probability, all elements in the sample space must be equally likely.

PROBLEM

A jar contains 5 red marbles, 8 green marbles, 10 blue marbles, and 4 white marbles. If I choose a marble at random what is the probability that it is:

(A) green! (B) red (C) not white

SOLUTION

(A) $\frac{8}{27}$

(B) $\frac{5}{27}$

(C) $\frac{23}{27}$

PROBLEM

A month of the year is chosen a random. What is the probability that it contains the letter y?

SOLUTION

Of the 12 months, January, February, May, and July contain the letter y. The probability is $\frac{4}{12} = \frac{1}{3}$.

PROBLEM

Tickets numbered 1 to 50 are mixed up and a ticket chosen at random. What is the probability that the ticket drawn has a number divisible by 4?

SOLUTION

The numbers divisible by 4 are 4, 8, 12, …, 48. There are 12 of them so the probability is $\frac{12}{50} = \frac{6}{25}$.

PROBLEM

A librarian shelves 9 books, one of which is an algebra book and one of which is a history book. What is the probability that both the algebra and history books are on the ends?

SOLUTION

There are 9! ways of arranging 9 books.

If the algebra book goes first and the history book goes last, there are 7! ways of arranging the other 7 books.

If the history book goes first and the algebra book goes last, there are 7! ways of arranging the other 7 books.

So the probability is

$$\frac{7!}{9!} + \frac{7!}{9!} = \frac{7 \cdot 6 \cdot 5 ... 1}{9 \cdot 8 \cdot 7 \cdot 6 \cdot 5 ... 1} + \frac{7 \cdot 6 \cdot 5 ... 1}{9 \cdot 8 \cdot 7 \cdot 6 \cdot 5 ... 1} = \frac{1}{72} + \frac{1}{72} = \frac{1}{36}$$

PROBLEM

If I draw two cards from a deck of 52 cards, what is the probability that both are aces?

SOLUTION

There are $_{52}C_2 = \frac{52!}{2! \cdot 50!} = \frac{52 \cdot 51 \cdot 50 \cdot 49 \cdot \ldots \cdot 3 \cdot 2 \cdot 1}{2 \cdot 1 \cdot 50 \cdot 49 \cdot \ldots \cdot 3 \cdot 2 \cdot 1} = \frac{52 \cdot 51}{2 \cdot 1} = 1{,}326$ ways of choosing 2 cards.

To get 2 aces, there are $_4C_2 = \frac{4!}{2! \cdot 2!} = \frac{4 \cdot 3 \cdot 2 \cdot 1}{2 \cdot 1 \cdot 2 \cdot 1} = 6$ combinations:

1) clubs, hearts, 2) clubs, diamonds, 3) clubs, spades,

4) hearts, diamonds, 5) hearts, spades, 6) diamonds, spades.

So the probability of 2 aces is $\frac{6}{1{,}326} = \frac{1}{221}$.

PROBLEM

A fair coin is tossed 3 times. What is the probability that 2 or more heads would show after the toss?

SOLUTION

Generate a sample space of all possibilities:

Toss 1	H	H	H	H	T	T	T	T
Toss 2	H	H	T	T	H	H	T	T
Toss 3	H	T	H	T	H	T	H	T

There are 8 possibilities and 4 of them show 2 or more heads.

So the probability is $\frac{4}{8} = \frac{1}{2}$

We can also do this without generating the sample space. There are $2^3 = 8$ possibilities and $_3C_2$ ways to get 2 heads and $_3C_3$ to get 3 heads. $_3C_2 = \frac{3!}{2! \cdot 1!} = 3$ and $_3C_3 = \frac{3!}{3! \cdot 0!} = 1$. So there are 4 ways to get 2 or more heads and its probability is $\frac{4}{8}=\frac{1}{2}$. (This method is easier when there are a large number of coins being tossed making the sample space too cumbersome to create.)

ODDS

Probability and odds have the same meaning—the chance of a random event occurring. However, they are not expressed in the same way. While the probability of an event is the ratio of the number of successful ways the event can occur compared with the *total number* of ways the event can occur, the odds of an event occurring is the ratio that compares the number of ways the event can occur with the number of ways the event cannot occur:

The odds in favor—the ratio of the number of ways that an outcome can occur compared to how many ways it cannot occur:

Number of successes: Number of failures.

The odds against—the ratio of the number of ways that an outcome cannot occur compared to in how many ways it can occur:

Number of failures: Number of successes.

Example:

If I roll a die, the probability of rolling a 3 are $\frac{1}{6}$. Rolling a 3 is a success, while rolling a 1, 2, 4, 5, 6 are failures. So the odds in favor of rolling a 3 are 1 to 5. The odds against rolling a 3 is 5 to 1.

PROBLEM

The odds against being chosen for a committee are 7 to 2. What is the probability of being chosen for the committee?

SOLUTION

Since this is "odds against," there are 7 failures and 2 successes. So the probability of being chosen for the committee is $\frac{2}{2+7}=\frac{2}{9}$.

MUTUALLY EXCLUSIVE AND COMPLEMENTARY EVENTS

Events that are mutually exclusive (also called disjoint) are events that cannot happen at the same time. Examples of choosing mutually exclusive events are:

- Choose one student in a group. You either choose a boy or a girl.
- Turn on a lamp. It either turns on or it doesn't.
- View the results of your favorite baseball team. The team either won or lost.
- You and a friend buy a raffle ticket. You and your friend cannot both win. Note that it is possible that neither of you will win, but you and your friend both winning are still disjoint events.

Two events are described as complementary if they are the only two possible outcomes. The first three examples above are complementary events. For any event A, the probability of A complement is given by $P(A^C) = 1 - P(A)$.

Example:

- If the probability of choosing a boy is 62%, the probability of choosing a girl is $1 - 0.62 = 38\%$.
- If the probability that a lamp turns on is 98%, the probability that it does not turn on is $1 - 0.98 = 2\%$.
- If the probability that a baseball team wins a game is said to be 45%; the probability that the team loses is $1 - 0.45 = 55\%$.
- If the probability that you win a raffle is 1%, the probability that your friend wins is not 99%. That is because although you and your friend winning are mutually exclusive events (you both cannot win at the same time), they are not complementary. That is because other people are also entered into the raffle as well.

If two events A and B are mutually exclusive, the probability of A or B = $\text{Prob}(A) + \text{Prob}(B)$.

Example:

- If a cooler contains 5 Cokes, 8 Pepsis, 10 Sprites, and 7 waters, the probability of choosing a Coke or a Pepsi = $\frac{5}{30}+\frac{8}{30}=\frac{13}{30}$. The probability of not choosing water is $1-\frac{7}{30}=\frac{23}{30}$. That is because choosing water and not choosing water are complementary events.
- People were asked what their favorite pet was. 69 said dogs, 53 said cats, 21 said fish, 5 said "other." 32 said they did not like pets. The probability of choosing someone who preferred a dog or a cat = $\frac{69}{180}+\frac{53}{180}=\frac{122}{180}=\frac{61}{90}$. Note that if the question asked the probability of choosing someone who *liked* dogs or cats, we could not answer. *Preferring* a dog or cat are mutually exclusive events—you either prefer one or the other. But *liking* dogs and cats are not mutually exclusive events as it is possible to like both.

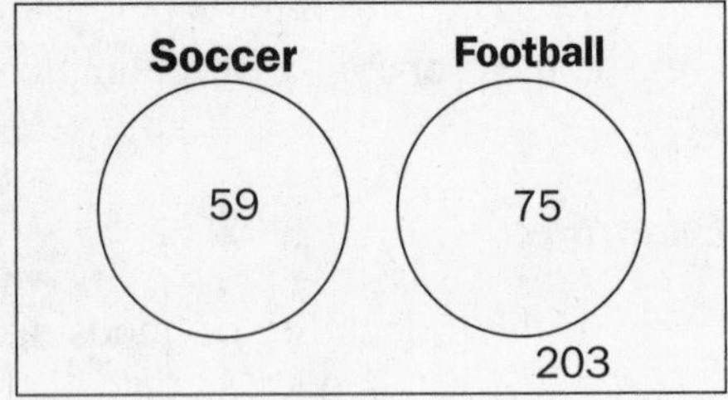

Probability questions can be answered by examining a Venn diagram (see Chapter 8 for more about Venn diagrams). In a school, students can only be in one fall sport. This Venn diagram shows how many students participate in soccer or football. The probability that a student is chosen who plays soccer or football = $\frac{59+75}{59+75+203}=\frac{134}{337}=39.8\%$. The probability of choosing a student who does not play soccer is $1-\frac{59}{337}=\frac{278}{337}$ = 82.5% as playing soccer and not playing soccer are complementary events.

NON-MUTUALLY EXCLUSIVE EVENTS

If two events *A* and *B* are *not* mutually exclusive (meaning that they *can* occur at the same time), we use the general formula to find the probability of A or B occurring. The word "or" means "A or B" or both.

Probability of A or B occurring = Prob(A) + Prob(B) − Prob(both A and B occurring)

Problems asking for this probability are either expressed in a paragraph, a Venn diagram, or a table.

Example:

- In a class of 25 students, 15 are seniors, 12 have brown hair, and 8 are seniors with brown hair. We want the probability of choosing a senior *or* a student with brown hair. Note that being a senior and having brown hair are *not disjoint* events – they *can* happen at the same time. So

  Prob(senior or brown hair) = Prob(senior) + Prob(brown hair) − Prob(senior with brown hair)

  $$\text{Prob(senior or brown hair)} = \frac{15}{25} + \frac{12}{25} - \frac{8}{25} = \frac{19}{25} = 76\%.$$

- A poll was taken as to how people got to work. The result is shown in the Venn diagram. To find the probability that someone takes a car *or* a train (which are *not* mutually exclusive events), the easiest way is finding $\frac{120+15+55}{120+15+55+10} = \frac{190}{200} = 95\%$. This is not a violation of the general formula as there are 135 people who drive a car, 70 people who take the train, and 15 people who do both. $\frac{135}{200} + \frac{70}{200} - \frac{15}{200} = \frac{190}{200} = 95\%$.

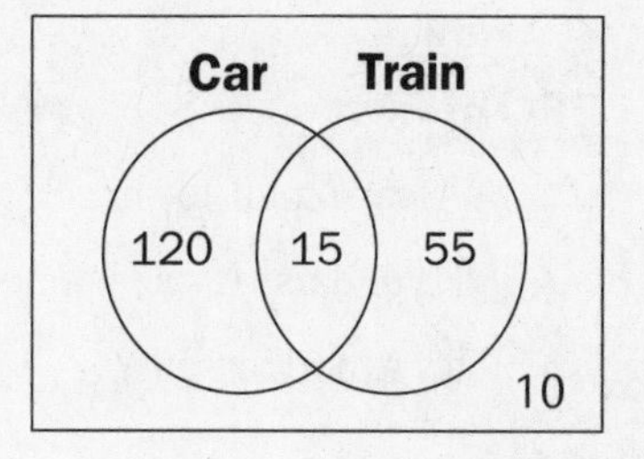

- The pass/fail status of 30 students is shown in the following table:

	Passing	Failing	Total
Boy	15	3	18
Girl	7	5	12
Total	22	8	30

The probability of choosing a student who is a boy *and* passing is $\frac{15}{30}=\frac{1}{2}=50\%$. The probability of choosing a student who is a boy *or* passing is found by either $\frac{15+7+3}{30}=\frac{25}{30}=\frac{5}{6}=83.3\%$ or $\frac{22+18-15}{30}$ $=\frac{25}{30}=\frac{5}{6}=83.3\%$.

PROBLEM

At a Florida road intersection, it is shown that 48% of the drivers go straight, 26% of the drivers turn right, 23% of the drivers turn left, and 3% of the drivers make U-turns. What is the probability that a driver turns right or left?

SOLUTION

Since these events are mutually exclusive, the probability that a driver turns right or left = 26% + 23% = 49%.

PROBLEM

A car rental lot has 35 Chevrolets, 44 Fords, 75 Toyotas, 68 Nissans, and 45 Hyundais. A car is chosen at random. What is the probability that the car is an American car?

SOLUTION

Since these events are mutually exclusive, the probability is that the car is American.

$$=\frac{35+44}{267}=\frac{79}{267}=29.6$$

PROBLEM

A number is chosen from 1 to 20. What is the probability that it is divisible by 2 or divisible by 3?

SOLUTION

Divisible by 2: 2, 4, 6, 8, 10, 12, 14, 16, 18, 20

Divisible by 3: 3, 6, 9, 12, 15, 18

There are 10 numbers divisible by 2, 6 numbers divisible by 3, but 6, 12, and 18 are counted twice.

So there are 13 numbers divisible by 2 or 3 and the probability is $\frac{13}{20}$.

PROBLEM

During a particular week, a doctor saw 125 patients. He treated 75 people with high blood pressure, 62 people with arthritis, and 41 people who had both. If a patient is chosen at random, what is the probability that the patient has arthritis or high blood pressure?

SOLUTION

Prob(arthritis or high BP) = Prob(arthritis) + Prob(high BP) − Prob(both)

$$\text{Prob(arthritis or high BP)} = \frac{75}{125} + \frac{62}{125} - \frac{41}{125} = \frac{96}{125}$$

PROBLEM

An Internet poll asks people whether they ever took a cruise, went on a bus tour, or both. The results are shown in the Venn diagram.

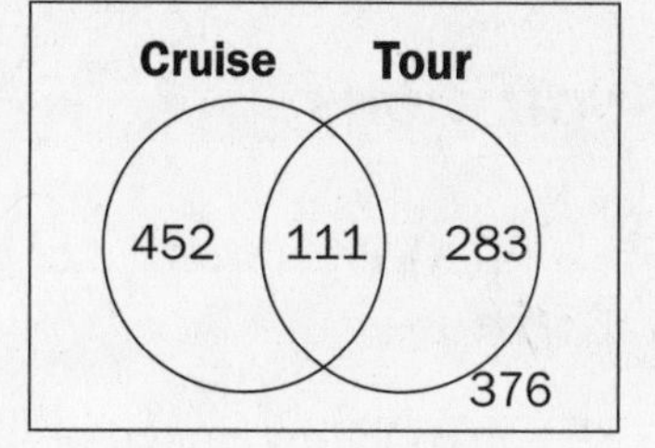

Find the probability of a person who went

(A) on a cruise only

(B) on a bus tour only

(C) on both a cruise and a bus tour

(D) on a cruise or a bus tour

(E) on a cruise or a bus tour, but not both

SOLUTIONS

(A) $\text{Prob(cruise only)} = \frac{452}{1,222}$

(B) $\text{Prob(bus tour only)} = \frac{283}{1,222}$

(C) $\text{Prob(both)} = \dfrac{111}{1,222}$

(D) $\text{Prob (cruise or bus tour only)} = \dfrac{452+111+283}{1,222} = \dfrac{846}{1,222}$

(E) $\text{Prob (cruise or bus tour but not both)} = \dfrac{452+283}{1,222} = \dfrac{735}{1,222}$

PROBLEM

The chart shows the on-time record of two airlines.

	On-time	Late
West Jet	52	75
United	38	60

Find the probability of a flight being from

(A) West Jet or on-time

(B) United Airlines or late.

SOLUTIONS

(A) $\text{Prob(West Jet or on-time)} = \dfrac{52+75+38}{225} = \dfrac{165}{225} = \dfrac{11}{15}$

(B) $\text{Prob (United or late)} = \dfrac{38+60+75}{225} = \dfrac{173}{225}$

CONDITIONAL PROBABILITY

In Florida, it is more likely to have rain if it is summer, rather than winter. The probability of rain in Florida is dependent on the month of the year. This is called conditional probability. Conditional probability is the probability event A occurs, given that event B occurs. We write this as Prob (A|B).

Examples of conditional probability include:

Finding the probability that a student is on the honor roll, given that he plays a school sport.

Finding the probability that a student is absent, given that it is a Friday.

Finding the probability that a man watches basketball on TV, given that he plays basketball.

Finding the probability that a woman exercises, given that she is married.

Finding the probability that if a person goes to a seafood restaurant, he will order seafood.

Example:

Suppose we have a class composed of the following:

	Senior	Junior	Total
Boy	8	4	12
Girl	3	5	8
Total	11	9	20

A student is chosen at random.

The probability of choosing a boy is $\frac{12}{20}=\frac{3}{5}=60\%$.

The probability of choosing a senior is $\frac{11}{20}=55\%$.

The probability of choosing a boy and a senior is $\frac{8}{20}=\frac{2}{5}=40\%$

The probability of choosing a boy or a senior is $\frac{8+4+3}{20}=\frac{15}{20}=\frac{3}{4}=75\%$.

Note that all of these have a denominator of 20.

Suppose we wish the probability of choosing a senior *given* that we have chosen a boy. This is an example of conditional probability. The condition is choosing a boy. Since there are 12 boys and 8 of them are seniors, the denominator is no longer 20 but 12. So the probability of choosing a senior given that we have chosen a boy is Prob (Senior|Boy) $=\frac{8}{12}=\frac{2}{3}=66.\bar{6}\%$. Since, as we saw above, the probability of choosing a senior is 55%, it is more likely to choose a senior, given that we choose a boy.

If we want the probability of choosing a boy *given* that we have chosen a senior, we see that there are 11 seniors and 8 of them are boys, so $\text{Prob(Boy|Senior)} = \frac{8}{11} = 72.7\%$.

PROBLEMS

	Senior	Junior	Total
Boy	8	4	12
Girl	3	5	8
Total	11	9	20

Using the chart, find

(A) the probability of choosing a girl given that we have chosen a junior.

(B) the probability of choosing a junior, given that we have chosen a girl.

SOLUTIONS

(A) $\text{Prob(Girl|Junior)} = \frac{5}{9}$

(B) $\text{Prob(JuniorGirl)} = \frac{5}{8}$

INDEPENDENT EVENTS

Two events are independent if knowing that one occurring does not change the probability of the other occurring. Flipping coins are independent events. If the first 5 flips are heads, the 6th coin has a 50% probability of being heads, assuming the coin is fair.

The weather in California today is independent of the weather in New York. If it is sunny in California, we are not willing to predict the weather in New York. However, the weather in Miami and the weather in Fort Lauderdale are not independent. Since the cities are only 20 miles apart, if it is raining in Miami, there is a good chance that it also will be raining in Fort Lauderdale.

Independent events are not the same as disjoint events. Disjoint events cannot happen at the same time. Independent events are events whose occurrences have no influence on each other.

To determine whether two events are independent we check to see whether Prob (A) · Prob (B) = Prob (both A and B occurring). If that is true, the events are independent.

Alternately, if Prob ($A|B$) = Prob (A), the events are independent. If the probability of A given B is the same as the probability of A, then B has no impact on A, which is the definition of independence.

Example:

Using the chart, determine whether choosing a boy and choosing a senior are independent events.

	Senior	**Junior**	**Total**
Boy	8	4	12
Girl	3	5	8
Total	11	9	20

Prob (boy) Prob (senior) $\stackrel{?}{=}$ Prob (senior boy)

$$\frac{12}{20} \cdot \frac{11}{20} \stackrel{?}{=} \frac{8}{20} \qquad 0.6 \cdot 0.55 \stackrel{?}{=} 0.4 \qquad 0.33 \neq 0.4$$

Since 0.33 does not equal 0.4, choosing a boy and choosing a senior are not independent. This goes hand-in-hand with what we previously showed. It was more likely to choose a senior if we had chosen a boy. Gender makes a difference.

Alternately, we can check whether Prob (Boy|Senior) = Prob (Boy). Since $\frac{8}{11} \neq \frac{12}{20}$, choosing a boy and choosing a senior are not independent.

Determine whether choosing a girl and choosing a junior are independent events.

$$\text{Prob (girl)} \cdot \text{Prob (junior)} \stackrel{?}{=} \text{Prob (junior girl)}$$

$$\frac{8}{20} \cdot \frac{9}{20} \stackrel{?}{=} \frac{5}{20} \quad 0.4 \cdot 0.45 \stackrel{?}{=} 0.25 \quad 0.18 \neq 0.25$$

Since 0.18 does not equal 0.25, choosing a girl and choosing a junior are not independent. It is more likely to choose a junior if we had chosen a girl. Gender makes a difference.

PROBLEMS

In an office building, both lawyers and doctors rent offices. Some smoke and some do not smoke. A person is chosen at random.

	Senior	Junior
Lawyer	24	38
Doctor	6	32

Using the chart, find the following probabilities:

(A) choosing a lawyer

(B) choosing a doctor

(C) choosing a smoker

(D) choosing a non-smoker

(E) choosing a lawyer who smokes

(F) choosing a doctor who does not smoke

(G) choosing a lawyer or a smoker

(H) choosing a doctor or a non-smoker

(I) choosing a lawyer, given that you choose a smoker

(J) choosing a doctor, given that you choose a non-smoker

(K) choosing a smoker, given that you choose a lawyer

(L) choosing a non-smoker, given that you choose a doctor

(M) Are choosing a lawyer and choosing a doctor mutually exclusive?

(N) Are choosing a lawyer and choosing a smoker mutually exclusive?

(O) Determine whether choosing a lawyer and choosing a smoker are independent events.

SOLUTIONS

(A) $\text{Prob}(\text{lawyer}) = \frac{62}{100} = 62\%$

(B) $\text{Prob}(\text{doctor}) = \frac{38}{100} = 38\%$

(C) $\text{Prob}(\text{smoker}) = \frac{30}{100} = 30\%$

(D) $\text{Prob}(\text{non-smoker}) = \frac{70}{100} = 70\%$

(E) $\text{Prob}(\text{lawyer who smokes}) = \frac{24}{100} = 24\%$

(F) $\text{Prob}(\text{doctor who does not smoke}) = \frac{32}{100} = 32\%$

(G) $\text{Prob}(\text{lawyer or smoker}) = \frac{24+38+6}{100} = \frac{68}{100} = 68\%$

(H) $\text{Prob}(\text{doctor or non-smoker}) = \frac{6+32+38}{100} = \frac{76}{100} = 76\%$

(I) $\text{Prob lawyer}|\text{smoker} \quad \frac{24}{30} \quad \frac{4}{5} \quad 80\%$

(J) $\text{Prob doctor}|\text{non-smoker} = \frac{32}{70} = \frac{16}{35} = 45.7\%$

(K) $\text{Prob}(\text{smoker}|\text{lawyer}) = \frac{24}{62} = \frac{12}{31} = 38.7\%$

(L) $\text{Prob}(\text{non-smoker}|\text{doctor}) = \frac{32}{38} = \frac{16}{19} = 84.2\%$

(M) Yes. In this building, you cannot be a lawyer and doctor at the same time.

(N) No. In this building you can be a lawyer and a smoker at the same time.

(O) No. Prob (lawyer) · Prob (smoker) = 0.62(0.3) = 0.186
Prob (lawyer smoker) = 0.24

EXPECTED VALUE

When you flip a coin, you expect that the percentage of heads will be 50% and the percentage of tails will be 50%. Rarely will they be exactly 50%, especially if you flip the coin many times. We call this an expected value—what you expect will happen.

If an experiment has possible numerical outcomes $x_1, x_2, \ldots x_n$ each having probability $P_1, P_2, \ldots P_n$, and each mutually exclusive, the expected value will be $x_1 p_1, x_2 p_2 \ldots x_n p_n$. Using summation notation, we can express this as $\sum_{i=1}^{n} x_i p_i$. Expected value can also be thought of as an average value.

Example:

A drug is administered to sets of three patients. Over a period of time, it is determined that the probability of 3 cures, 2 cures, 1 cure, and no cures are .70, .20, .09, and .01 respectively. The expected number of cures that can be expected in a group of three is calculated:

$$3(0.70) + 2(0.20) + 1(0.09) + 0(0.01) = 2.59.$$

When we give this drug to 3 patients, we expect that 2.59 will be cured.

Note that we cannot get 2.59 cures. We get 2 cures or 3 cures. But the expected value is an average, just as saying the average family in a community has 2.4 children.

PROBLEM

When playing the Pick-3 lottery, you have to choose a 3-digit number. If you win, you get \$500. If you lose, you get nothing. A lottery ticket costs \$1. We wish to find the expected value of this lottery game. What should your expectation be when you buy a lottery ticket?

SOLUTION

There are only two results, winning and losing. Numerically, that means \$500 or \$0.

By the multiplication principle, there are $10 \cdot 10 \cdot 10 = 1{,}000$ possibilities so the probability of winning is $\frac{1}{1{,}000}$ and the probability of losing is $\frac{999}{1{,}000}$.

So the expected value of the lottery is $500\ \frac{1}{1{,}000} + 0\ \frac{999}{1{,}000} = \frac{1}{2} = 0.50$. So you will win 50 cents on average when you buy a lottery ticket. But since the lottery ticket costs \$1, you will lose 50 cents on average every time you play the lottery. You can win the lottery but play it a lot of times and your expectation is losing 50 cents for every ticket you buy.

PROBLEM

In a community, the probability that a family will have a given number of cars (vehicles) is given by the chart below. What is the expected number of cars for a family in this community?

Cars	**0**	**1**	**2**	**3**	**4**
Probability	0.18	0.24	0.35	0.19	0.04

SOLUTION

Expected value $= 0(0.18) + 1(0.24) + 2(0.35) + 3(0.19) + 4(0.04) = 1.67$

The average family in the community owns 1.67 cars.

PROBLEM

Dollar-Dog-Night is a promotion that a professional baseball team runs several times during the season. They charge \$1.00 for a hot dog and hope that people will eat more of them than they would if prices were normal. The chart below gives the probability that a person will purchase the indicated number of hot dogs.

Hot dogs	0	1	2	3	4	5
Probability	0.32	0.28	0.20	0.15	0.03	0.02

(A) Find the expected number of hot dogs that a fan at the game will eat.

(B) If the baseball team makes a profit of 40 cents per hot dog and 25,000 fans attend the game, how much profit will the team make?

SOLUTION

(A) Expected value $= 0(0.32) + 1(0.28) + 2(0.20) + 3(0.15) + 4(0.03) + 5(0.02) = 1.35$
The average fan eats 1.35 hot dogs.

(B) The profit for 25,000 fans $= 25{,}000(1.35)(0.40) = \$13{,}500$

Drill Questions

1. All liberal arts students at Newton College must successfully complete one course taken from each of the following categories in order to graduate. In how many ways can a student satisfy the college requirement?

Science & Technology	Analysis	The Arts
Math	Political Science	Music
Physics	Philosophy	Art
Chemistry	Sociology	Photography
Astronomy	Logic	
Geology		
Programming		

(A) 13
(B) 30
(C) 42
(D) 72

2. Nine horses are in the Kentucky Derby. You wish to bet on the Trifecta. You need to choose correctly predicting which horse will win, will come in 2nd, and will come in 3rd. How many ways are there to do this?

(A) 24
(B) 504
(C) 729
(D) 362,880

3. In a meeting of 12 businesspeople, each shakes hands with the other. How many handshakes will there be?

(A) 24
(B) 66
(C) 132
(D) 12!

4. The letters in the word HEART are scrambled. What is the probability that the "H" will be first and the "T" will be last in the scrambled word?

(A) $\frac{1}{20}$

(B) $\frac{1}{6}$

(C) $\frac{1}{40}$

(D) $\frac{2}{5}$

5. A small class contains 3 boys and 6 girls. Their teacher chooses 3 students at random to serve on a committee. If Brian and Janice are boyfriend and girlfriend, what is the probability that they will both be chosen to be on the committee?

(A) $\frac{2}{9}$

(B) $\frac{1}{3}$

(C) $\frac{1}{84}$

(D) $\frac{1}{12}$

6. A study is done at a school on student lateness. The statistics taken on one day are shown in the table below. What is the probability that a student chosen either drove himself or herself or was late?

	On-time	Late
Drove themselves	54	8
Were driven	24	2
Took school bus	40	0

(A) 0.50
(B) 0.5625
(C) 0.0625
(D) 0.125

7. A study was done to determine whether people who purchase a particular brand of computer are more apt to switch to another brand of computer. The table below shows the results of the study for 100 people who originally purchased PC or Mac computers.

	Stayed with computer	Switched computers	Total
Started with PC	44	20	64
Started with Mac	30	6	36
Total	74	26	100

Which statements are true?

I. The probability that a person who originally purchases a PC is greater than the probability that a person who originally purchases a Mac.

II. The probability that a person who purchases a PC and switches computers is greater than the probability than a person who purchases a Mac and switches computers.

III. Purchasing a PC and switching computers are independent events.

(A) I and II only
(B) I and III only
(C) II and III only
(D) I, II, and III

8. Frank is sorting through his Christmas lights. Some are defective. The chart below shows a summary of the classification of the lights.

	Defective	OK	Total
Red	2	6	8
Green	5	7	12
Total	7	13	20

Frank chooses a bulb at random. Arrange the following probabilities from highest to smallest.

I. Probability that he chooses a red bulb.

II. If he chooses a red bulb, it is defective.

III. If he chooses a defective bulb, it is red.

(A) I – II – III
(B) I – III – II
(C) II – III – I
(D) III – II – I

9. Suppose choosing a new car color is independent of choosing a car brand. If the probability of choosing a Toyota is 15% and the probability of choosing a white Toyota is 3%, find the probability of choosing a white car.

(A) 0.45%
(B) 16.7%
(C) 18%
(D) 20%

10. A charity is raffling off a 60-inch TV worth $2,000; two 32-inch TV's worth $500 each; and 5 iPods worth $230 each. A raffle ticket costs $10 and 1,000 tickets are sold. What is your mean expectation for the raffle?

(A) Loses $7.27
(B) Loses $5.85
(C) Wins $2.73
(D) Wins $4.15

Answers to Drill Questions

1. **(D)** 6(4)(3) = 72

2. **(B)** 9(8)(7) = 504

3. **(B)** This is a combination problem as A shaking hands with B is the same as B shaking hands with A.

   There are ${}_{12}C_2 = \frac{12!}{2!\cdot 10!} = \frac{12\cdot 11\cdot 10\cdot 9\cdot \ldots \cdot 3\cdot 2\cdot 1}{2\cdot 1\cdot 10\cdot 9\cdot \ldots \cdot 3\cdot 2\cdot 1} = \frac{12\cdot 11}{2\cdot 1} = 66$ handshakes.

4. **(A)** This is a permutation problem as order counts.
   There are 5! = 120 ways to scramble the letters.

   If the H and T are in their proper positions, there are 3! ways to scramble the other 3 letters.

   So the probability that the H and the T will be in their proper positions is $\frac{3!}{5!} = \frac{3\cdot 2\cdot 1}{5\cdot 4\cdot 3\cdot 2\cdot 1} = \frac{1}{20}$.

5. **(D)** This is a combination problem as order doesn't count.

   There are ${}_{9}C_3 = \frac{9!}{3!\cdot 6!} = \frac{9\cdot 8\cdot 7\cdot 6\cdot 5\cdot 4\cdot 3\cdot 2\cdot 1}{3\cdot 2\cdot 1\cdot 6\cdot 5\cdot 4\cdot 3\cdot 2\cdot 1} = \frac{9\cdot 8\cdot 7}{3\cdot 2\cdot 1} = 84$ ways to choose the committee.

   If Brian and Janice are on the committee, there are 7 other people who can also be on the committee.

   So the probability that Brian and Janice will be on the committee is $\frac{7}{84} = \frac{1}{12}$.

6. **(A)** Total students: 54 + 24 + 40 + 8 + 2 + 0 = 128

   Students who drove themselves or were late: 54 + 8 + 2 = 64

   Probability that students drove themselves or were late: $\frac{64}{128} = 0.50$

7. **(A)** I. True: Prob (purchased PC) $= \frac{64}{100} = 0.64$

   Prob (purchased Mac) $= \frac{36}{100} = 0.36$

II. True: Prob (Switch|Mac) $= \frac{20}{64} = 0.3125$

Prob (swith| Mac) $= \frac{6}{36} = 0.1667$

III. False: Prob (purchased PC) = 0.64

Prob(Switch) $= \frac{26}{100} = 0.26$

Prob (purchased PC and switched) $= \frac{20}{100} = 0.20$

$0.64(0.26) = 0.1664 \neq 0.20$

8. **(B)** I. Prob(red) $= \frac{8}{20} = 0.4$

II. Prob (Defective |Rd) $= \frac{2}{8} = 0.25$

III. Prob (Red|Defective) $= \frac{2}{7} = 0.286$

9. **(D)** When independent, Prob (Toyota) · Prob (White Car) = Prob (White Toyota)

$0.15 \cdot \text{Prob(White Car)} = 0.03 \Rightarrow \text{Prob(White Car} = \frac{0.03}{0.15} = 0.2 = 20\%$

10. **(B)**

	60-inch TV	32-inch TV	Ipod	Nothing
X	2,000	500	230	0
Prob (X)	$\frac{1}{1,000}$	$\frac{2}{1,000}$	$\frac{5}{1,000}$	$\frac{992}{1,000}$

Expected Value (win)

$$= 2,000\left(\frac{1}{1,000}\right) + 500\left(\frac{2}{1,000}\right) + 230\left(\frac{5}{1,000}\right) + 0\left(\frac{992}{500}\right) = \frac{4,150}{1,000} = \$4.15$$

= 2,000

Person wins \$4.15 − \$10.00 = −\$5.85.

CHAPTER 5

Statistics and Data Analysis

CHAPTER 5

STATISTICS AND DATA ANALYSIS

MEASURES OF CENTRAL TENDENCY

Mean or average: The mean of a set of data is the average and is usually denoted $\bar{x}$. To find it, add up all the data and divide by the number of pieces of data.

Example:

8 people went out to dinner. The list below shows what they spent. Find the average cost.

\$18.55, \$21.35, \$17.45, \$20.50, \$24.25, \$14.75, \$19.65 and \$22.90.

$$\bar{x} = \frac{18.55 + 21.35 + 17.45 + 20.50 + 24.25 + 14.75 + 19.65 + 22.90}{8}$$

$$= \frac{159.40}{8} = \$19.93$$

PROBLEM

Find the average height of the starting 5 members of a basketball team whose heights are

6′6″, 6′10″, 5′11″, 7′1″, and 6′7″.

SOLUTION

To start, convert all heights to inches:

$$\bar{x} = \frac{78 + 82 + 71 + 85 + 79}{5} = \frac{395}{5} = 79$$

79 inches is equal to 6 feet 7 inches. Thus, the average height is 6 feet 7 inches.

PROBLEM

Sheldon averaged 88.5 over 4 rounds of golf. He shot 93 in his 2nd round, 84 in his 3rd round, and 90 in his 4th round. What was his first round score?

SOLUTION

Let x be his first round score:

$$\frac{x+93+84+90}{4} = 88.5 \Rightarrow x + 267 = 354 \Rightarrow x = 87$$

Median: The median of a set of data is the middle score. To find the median, first put the data in increasing or decreasing order. Let n represent the number of pieces of data. If n is odd, the median is the data value: $\frac{n+1}{2}$. If n is even, take the average of data $\frac{n}{2}$ and the piece of data directly after it. When one of the data is much larger or smaller than the rest of the data, the mean can be strongly affected, while the median is not.

Example:

A small class is made up of the students with first names: Susie, Joe, Michael, Shawn, Caroline, Steve, Kurt, Jennifer, Matthew, and Jonathan. Find the median name length.

The lengths of the names are 5, 3, 7, 5, 8, 5, 4, 8, 7, and 8. When put in order we get: 3, 4, 5, 5, **5**, 7, 7, 8, 8, 8. Since there are 10 pieces of data, the median is the average of the 5th and 6th pieces of data, which are 5 and 7. The median is 6.

PROBLEM

15 people entered a hot-dog eating contest. The number of hot dogs they ate are:

23, 14, 19, 21, 20, 12, 21, 24, 40, 19, 20, 23, 25, 12, and 15

(a) Find the mean and the median number of hot dogs eaten.

(b) If the person eating 40 hot dogs were disqualified, how would that affect the mean and the median?

SOLUTION

$$\bar{x} = \frac{23+14+19+21+20+12+21+24+40+19+20+23+25+12+15}{15}$$

$$= \frac{308}{15} = 20.5\overline{3}$$

Data sorted: 12, 12, 14, 15, 19, 19, 20, 20, 21, 21, 23, 23, 24, 25, 40.

The median is the 8th score which is 20.

If 40 is eliminated, $\bar{x} = \frac{308-40}{14} = \frac{268}{14} = 19.14$

The median will be the average of the 7th and 8th score which is still 20.

Mode: The mode of a set of data is the data that occurs the most often. Mode is rarely used.

Example:

In the class mentioned above, since a name with 5 letters and a name with 8 letters both occur three times, more than any other number of letters, that set of data has two modes: 5 and 8.

PROBLEM

Movie patrons were asked to rate the movie they had just seen. They were asked to rate it with a 1 for strongly disliking the movie and a 5 for strongly liking it. Using tally marks to track the data, the chart that follows shows the results:

5	~~IIII~~ ~~IIII~~ ~~IIII~~ III
4	~~IIII~~ ~~IIII~~ ~~IIII~~ IIII
3	~~IIII~~ ~~IIII~~ II
2	~~IIII~~ II
1	III

Find the average score, the median score, and the mode.

SOLUTION

Score	Frequency
5	18
4	19
3	12
2	7
1	3

Mean: $\bar{x} = \frac{18(5)+19(4)+12(3)+7(2)+3(1)}{59} = \frac{219}{59} = 3.71$

Median is the 30th score, which is 4.

Mode is the most common score, which is 4.

COMPARING THE MEAN, MEDIAN, AND MODE IN A VARIETY OF DISTRIBUTIONS

It is sometimes useful to make comparisons about the relative values of the mean, median, and mode. This section presents steps to do that without having to calculate the exact values of these measures.

Step 1: If a bar graph is not provided, sketch one from the information given in the problem.

The graph will either be skewed to the left, skewed to the right, or approximately normal. A **skewed** distribution has one of its tails longer than the other.

- A graph skewed to the left will look like Figure 1; its left tail is longer. The order of the three measures is mean < median < mode (alphabetical order).
- A graph skewed to the right will look like Figure 2; its right tail is longer. The order of the three measures is mode < median < mean (reverse alphabetical order).
- A graph that is approximately normal (also called bell-shaped) will look like Figure 3, and the mean = median = mode.

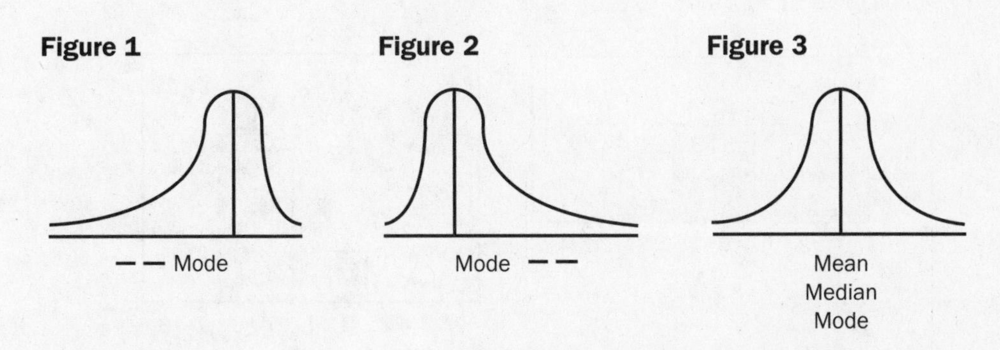

Step 2: Write the word "mode" under the highest column of the bar graph, because the mode is the most frequent. If the graph is skewed right or left, the positioning of the mode establishes the order for the remaining two terms according to the information provided above. If the graph is approximately symmetrical, the value of all three terms are approximately equivalent.

PROBLEM

On a trip to the Everglades, students tested the pH of the water at different sites. Most of the pH tests were at 6. A few read 7, and one read 8. Select the statement that is true about the distribution of the pH test results.

(A) The mode and the mean are the same.

(B) The mode is less than the mean.

(C) The median is greater than the mean.

(D) The median is less than the mode.

SOLUTION

Sketch a graph.

The graph is skewed to the right. The mode is furthest left. Thus, mode $<$ median $<$ mean.

Choices (A), (C), and (D) do not coincide with what has been established in terms of relative order. (B) is the only choice that does follow from our conclusions. Thus, (B) is the correct response.

PROBLEM

Estimate the median of each distribution shown below and describe how the mean compares to it.

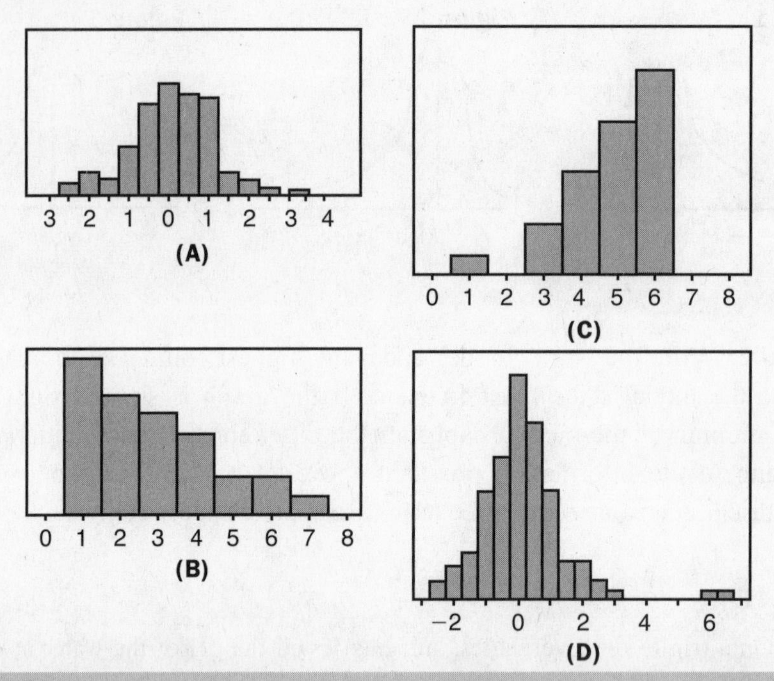

SOLUTION

Graph (A) is symmetric with a median at about 0. The mean will also be about 0, since it approximates a normal curve. Graph (B) is right-skewed with a median of about 3. The mean will be pulled more toward the values of $4-7$ than it will be toward $1-2$, so the mean will be greater than 3. Graph (C) is left-skewed with a median of about 5. The mean will be pulled toward the lower values, so it is less than 5. Graph (D) is symmetric with a few extreme high values. The median is about 0, and the mean will be just slightly larger due to the outliers around 6.

MEASURES OF VARIABILITY

While mean, median, and mode measure the center of a data set, they say nothing about how spread out the data is. We call this spread a measure of **variability**. A manufacturer of light bulbs would prefer small variability in the number of hours the bulbs will likely burn. A track coach who needs to decide which athletes go on to the finals may want larger variability in heat times because it will be easier to decide who are truly the fastest runners. There are two measures of variability: **range and standard deviation**.

Range: The range is the difference between the highest and lowest data values. Given all of the data, the range is easy to find. The larger the range, the greater the spread of the data.

Example:

If the heaviest person in a room is 205 pounds and the lightest person is 130 pounds, the range is 205 – 130 = 85 pounds.

Variance and Standard Deviation: The standard deviation is a measure of how spread out the data is from the mean and is quite useful for data that is fairly symmetric. The greater the standard deviation, the greater the spread of the data from the mean.

The variance tells us how much variability exists in a distribution. It is the "average" of the squared differences between the data values and the mean. The variance s^2 is calculated with the formula

$$s^2 = \frac{1}{n-1}\sum\left(x_i - \overline{x}\right)^2$$

where n is the number of data points, Σ refers to summation, x_i represents each data value, and $\overline{x}$ is the mean.

The formula for the standard deviation s is therefore $\sqrt{\text{variance}}$. Typically a calculator is needed to find the standard deviation of a data set. The standard deviation is used for most applications in statistics. It can be thought of as how far the typical observation lies from the mean.

Example:

Six movie reviewers rated a current movie on a star basis where 1 star is poor and 5 stars is excellent. If the ratings are

2, 5, 4, 5, 5, 3.

find the mean and the standard deviation.

The mean is $\overline{x} = \dfrac{2+5+4+5+5+3}{6} = \dfrac{24}{6} = 4$

The variance $s^2 = \dfrac{(2-4)^2+(5-4)^2+(4-4)^2+(5-4)^2+(5-4)^2+(3-4)^2}{5}$

$$s^2 = \frac{4+1+0+1+1+1}{5} = \frac{8}{5} = 1.6$$

The standard deviation $s = \sqrt{1.6} = 1.26$

Outliers: An outlier is a score that is dramatically different from the other scores in a group of data. Outliers can have a huge effect on the mean and standard deviation and it is important to know whether the outlier is a true score and not a typographical error, or whether it occurred because of special circumstances.

At this stage of statistics, the standard deviation by itself tells us little about how the data is distributed. However, when two or more data sets are compared, the standard deviations of each allow us to compare the data sets.

Example:

The chart on the next page shows exam scores of 4 classes of 5 students each. Each class has a mean of 80, but the data is quite different. The mean, median, range, and standard deviation of each class is also shown. Interpret the variability of the classes.

Class 1	Class 2	Class 3	Class 4
80	90	100	100
80	85	90	100
80	80	80	100
80	75	70	100
80	70	60	0
$\overline{x}$ = 80 Median = 80 Range = 0 St. Dev = 0	$\overline{x}$ = 80 Median = 80 Range = 20 St. Dev = 7.90	$\overline{x}$ = 80 Median = 80 Range = 40 St. Dev = 15.81	$\overline{x}$ = 80 Median = 100 Range = 100 St. Dev = 17.89

Class 1 has a standard deviation of 0 because there is no spread about the mean of 80. Classes 2, 3, and 4 have larger standard deviations meaning that there is a bigger spread about the mean. 0 is an outlier in class 4, which dramatically changes the mean and standard deviation.

PROBLEM

In a used car lot, the standard deviation of the car prices is $750. If the owner raises the price of every car by $250, what would be the standard deviation of the new car prices?

(A) $500

(B) $750

(C) $1,000

(D) Impossible to determine

SOLUTION

(B) Since the spread of the data does not change, the standard deviation remains the same. A $1,000 car will now cost $1,250 while a $10,000 car will now cost $10,250. The spread will stay the same at $9,000.

PROBLEM

In another used car lot, the standard deviation of the car prices is $750. If the owner raises the cost of every car by 10%, what would be the standard deviation of the new car prices?

(A) $450

(B) $750

(C) $825

(D) Impossible to determine

SOLUTION

(C) Multiplying each car price by 1.1 would increase the spread. A $1,000 car will now cost $1,100 while a $10,000 car will now $11,000. The spread went from $9,000 to $9,900.

Thus, the standard deviation would likewise increase by 10%.

PROBLEM

The effect of global warming has not been higher temperatures, but bigger extremes in cold temperatures and warm temperatures. If the temperatures of cities in the United States were averaged in the years 1960 (pre-global warming) and 2015 (during global warming), what would be the approximate effect on the mean and standard deviation?

Mean:	Standard Deviation:
______ Would go up	______ Would go up
______ Would stay the same	______ Would stay the same
______ Would go down	______ Would go down

SOLUTION

If some cities got colder and others got warmer, there would not be a big difference in the means of the two years. However, if some cities got colder and others warmer, the spread from the mean would be greater, so the standard deviation would go up.

PROBLEM

Six people are standing on a subway platform. If the average position on the platform is measured, arrange the following choices in order from smallest to largest standard deviation.

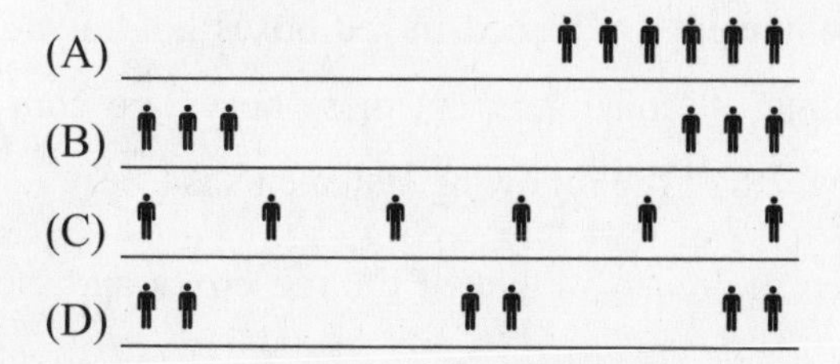

SOLUTION

(A), (C), (D), (B)

In (A) all the people are standing at the end of the platform so the standard deviation should be close to zero. In (B), (C), and (D), the mean position is in the center of the platform. Choice (B) has the people furthest from the center so it will have the largest standard deviation.

PROBLEM

The standard deviation of rainfall in Seattle, Washington, during the summer is 0.8 inches. The standard deviation of rainfall in Miami, Florida, is 1.6 inches. Which of the following statements is true?

(A) It rains twice as much in Miami as in Seattle.

(B) It rains twice as much in Seattle as in Miami.

(C) Miami is more likely than Seattle to have days with no rain and days with a lot of rain.

(D) There are more rainy days in Miami than in Seattle.

SOLUTION

(C) With a larger standard deviation, there is a greater spread in Miami's rainfall than in Seattle's. So Miami is more likely to have days with no rain and others with a lot of rain. Seattle is more likely to have many days with moderate rain. The correct answer is (C).

PROBLEM

A large conference is held in a hotel and lunch is provided. Lunch consists of an 8-ounce sandwich, a 4-ounce cup of juice, and a package of cookies weighing 3 ounces. They are placed in a styrofoam container weighing 1 ounce. These styrofoam containers are placed in red boxes holding 12 containers or blue boxes holding 20 containers. If these boxes are completely filled with containers, which of the following statements is true?

I. There is no difference between the mean weight of the red boxes and the blue boxes.

II. There is no difference between the standard deviation weight of the red boxes and the blue boxes.

(A) I only

(B) II only

(C) I and II

(D) Neither I nor II

SOLUTION

(B) All of the red boxes weigh 12 pounds and all of the blue boxes weigh 20 pounds. The average weight of the red boxes is 12 and the average weight of the blue boxes is 20. Therefore, statement I is not true. Since all the red boxes weigh the same, the standard deviation is zero and the same can be said of the blue boxes. The correct answer is (B).

NORMAL DISTRIBUTIONS

One of the most important distributions in statistics is called the **normal distribution**. If a histogram is "smoothed out," many times its curve will appear symmetric, single-peaked, and bell-shaped. These are called *normal curves* or the *bell curve*. On the next page is a graph of a normal curve. Note how most of the data is in the center and it tails off symmetrically to the sides with little data at the far left and right.

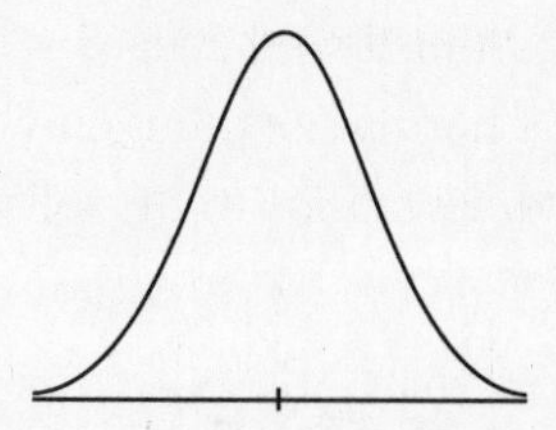

The reason these curves are so important in statistics is that so much of real-life statistics appears "normal." Here are some common examples:

- Heights: Most people are average height with fewer people being really short or really tall.
- Prices: Go to many markets and price a 2-liter bottle of the same brand of soda. The data will appear normal with most prices about average. Few prices will be cheaper and few will be more expensive.
- Wages: Most people earn an average amount of money while fewer people earn either a little money or a lot of money.
- Time it takes to get to work: Most days it might take 30 minutes to drive to work while fewer days it might take 20 minutes or 40 minutes.
- Grades: Collect GPAs for a class of students and many students will have an average GPA. Fewer will have either low GPAs or high GPAs.

The normal distribution follows an important rule called the **68-95-99.7% rule.** It states that in any normal distribution,

- 68% of the data lies within one standard deviation of the mean. In the figure below, we have a normal distribution with the mean $\overline{x}$ in the center. Going out one standard deviation to the left and right of $\overline{x}$ will represent 68% of the data.

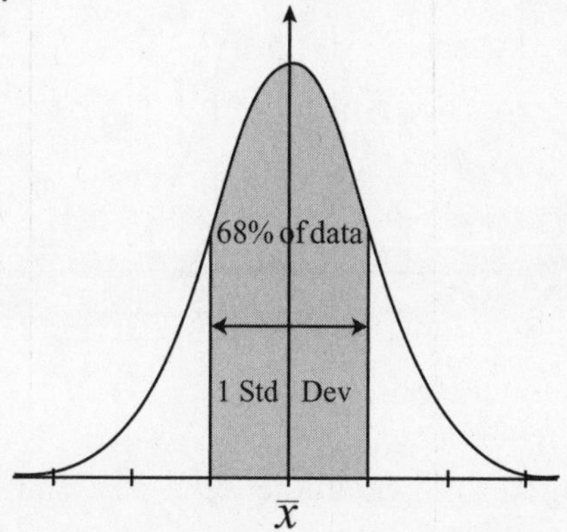

- 95% of the data lies within two standard deviations of the mean. In the figure below, we have a normal distribution with the mean $\overline{x}$ in the center. Going out two standard deviations to the left and right of $\overline{x}$ will represent 95% of the data.

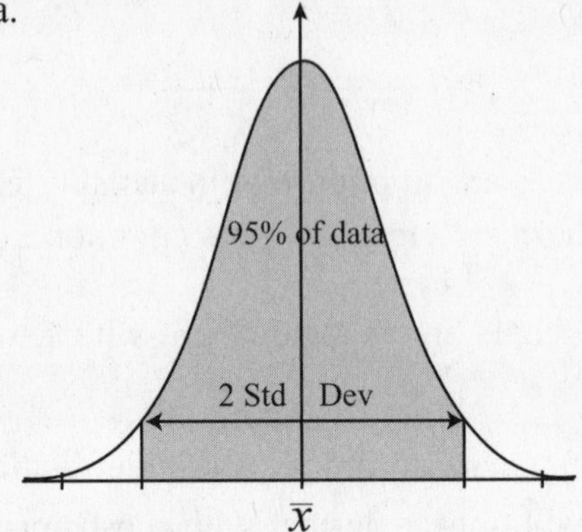

- 99.7% (or just about all) of the data lies within three standard deviations of the mean.

The more "normal" a distribution is, these relationships become closer to being perfectly true. No distribution is perfectly normal and, therefore, these relationships are approximations in most real-life settings.

Example:

The distribution of heights of adult American men is approximately normal with a mean of 69 inches and a standard deviation of 2.5 inches. On the normal curve below, we mark the values for the mean and 1, 2, and 3 standard deviations above and below the mean.

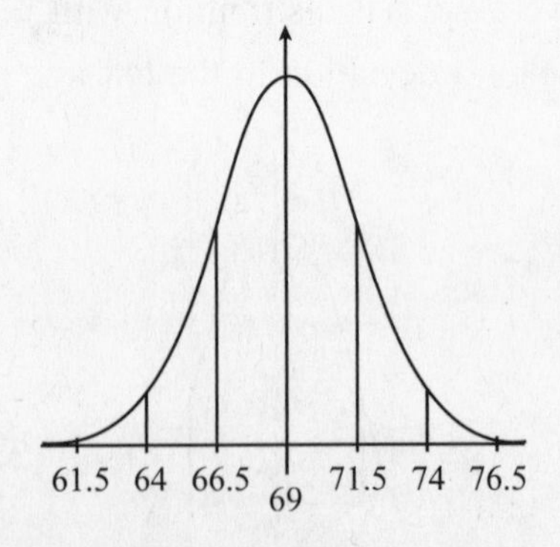

We can make the following observations:

- 68% of adult American men are between 66.5 and 71.5 inches (5'6.5" and 5'11.5") tall.

- 95% of adult American men are between 64 and 74 inches (5'4" and 6'2") tall.
- 99.7% of adult American men are between 61.5 and 76.5 inches (5'1.5" and 6'4.5") tall.

PROBLEM

The length of human pregnancies from conception to birth is approximately normally distributed with a mean of 284 days and a standard deviation of 11 days. Find the approximate percentage of pregnancies

(A) over 284 days

(B) over 295 days

(C) between 262 and 284 days

(D) less than 306 days

SOLUTION

The graph of the normal curve with standard deviation marks is shown below.

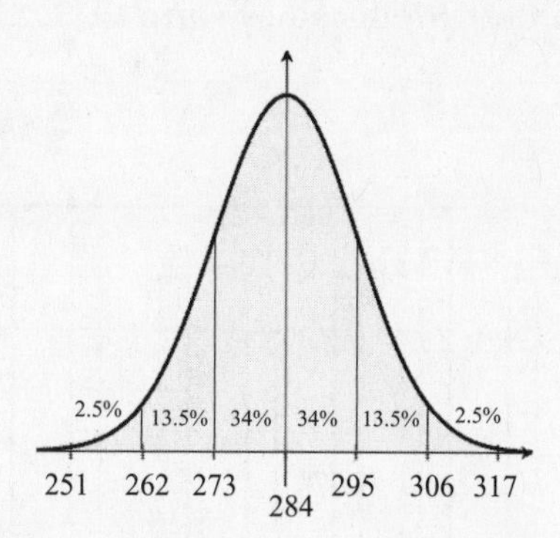

(A) Because of symmetry, 50% of pregnancies are greater than 284 days.

(B) 295 days is one standard deviation over the mean. 16% of pregnancies are greater than 295 days.

(C) 262 is two standard deviations below the mean. 47.5% of pregnancies are between 262 and 284 days.

(D) 306 is two standard deviations above the mean. 97.5% of pregnancies are below 306 days.

DATA ANALYSIS

Data analysis often involves putting numerical values into picture form, such as bar graphs, line graphs, and circle graphs. In this manner, we gain a more intuitive understanding of the given information.

BAR GRAPHS

Bar graphs are used to compare amounts of the same measurements. The following bar graph compares the number of bushels of wheat and corn produced on a farm from 1975 to 1985. The horizontal axis for a bar graph consists of categories (e.g., years, ethnicity, marital status) rather than values, and the widths of the bars are uniform. The emphasis is on the height of the bars. Contrast this with histograms, discussed next.

PROBLEM

According to the graph below, in which year was the least number of bushels of wheat produced?

Number of Bushels (to the Nearest 5 Bushels) of Wheat and Corn Produced by Farm RQS, 1975–1985

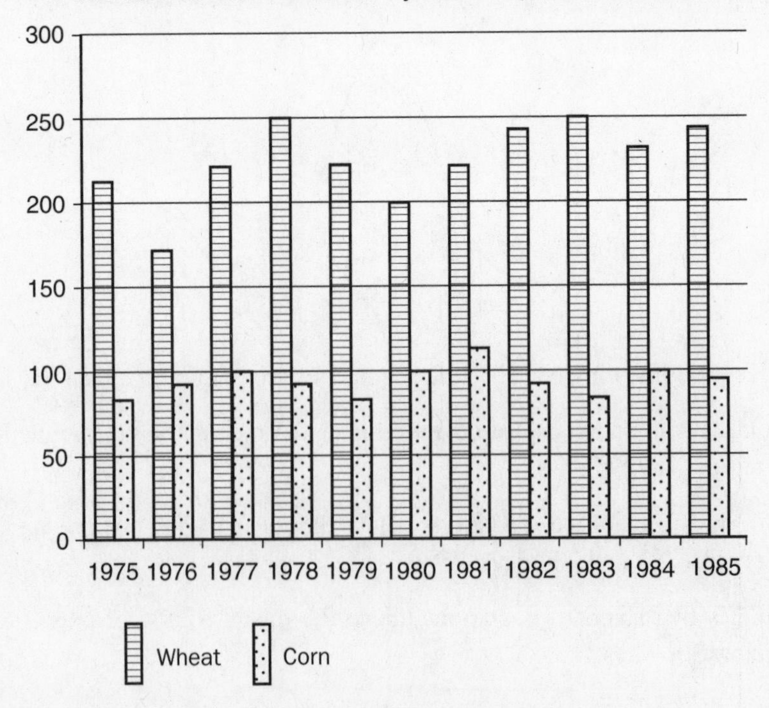

SOLUTION

By inspecting the graph, we find that the shortest bar representing wheat production is the one for 1976. Thus, the least number of bushels of wheat produced in 1975–1985 occurred in 1976.

HISTOGRAMS

A **histogram** is an appropriate display for quantitative data. It is used primarily for continuous data, but may be used for discrete data that have a wide spread. The horizontal axis is broken into intervals that do not have to be of uniform size. Histograms are also good for large data sets. The area of the bar denotes the value, not the height, as in a bar graph.

The histogram below shows the amount of money spent by passengers on a ship during a recent cruise to Alaska.

Example:

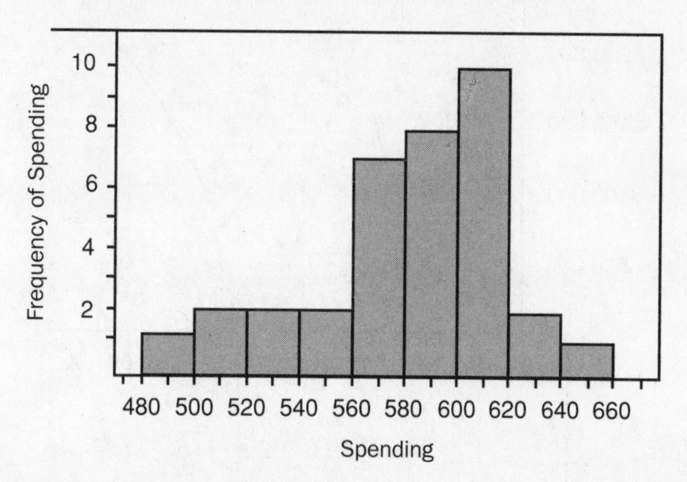

The intervals have widths of $20. One person spent between $480 and $500, two spent between $500 and $520, and so on. We cannot tell from the graph the precise amount each individual spent.

The distribution has a shape skewed to the left with a peak around $600 to $620. The data are centered at about $590.This is about where half of the observations will be to the left and half to the right. The range of the data is about

$180, but the clear majority of passengers spent between $560 and $620. There are no extreme values present or gaps within the data.

LINE GRAPHS

Line graphs are very useful in representing data on two different but related subjects. Line graphs are often used to track the changes or shifts in certain factors. In the next problem, a line graph is used to track the changes in the amount of scholarship money awarded to graduating seniors at a particular high school over the span of several years.

PROBLEM

According to the line graph below, how much did the scholarship money increase between 1987 and 1988?

Amount of Scholarship Money Awarded to Graduating Seniors, West High, 1981--1990

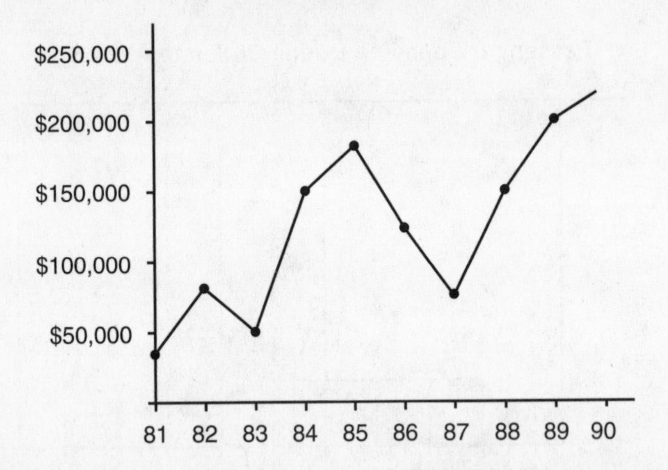

SOLUTION

To find the increase in scholarship money from 1987 to 1988, locate the amounts for 1987 and 1988. In 1987, the amount of scholarship money is halfway between $50,000 and $100,000, or $75,000. In 1988, the amount of scholarship money is $150,000. The increase is thus $150,000 − 75,000 = $75,000.

PIE CHARTS

Circle graphs (or **pie charts**) are used to show the breakdown of a whole picture. When the circle graph is used to demonstrate this breakdown in terms of percents, the whole figure represents 100% and the parts of the circle graph represent percentages of the total. When added together, these percentages add up to 100%. The circle graph in the next problem shows how a family's budget has been divided into different categories by using percentages.

PROBLEM

Using the budget shown below, a family with an income of $3,000 a month would plan to spend what amount on housing?

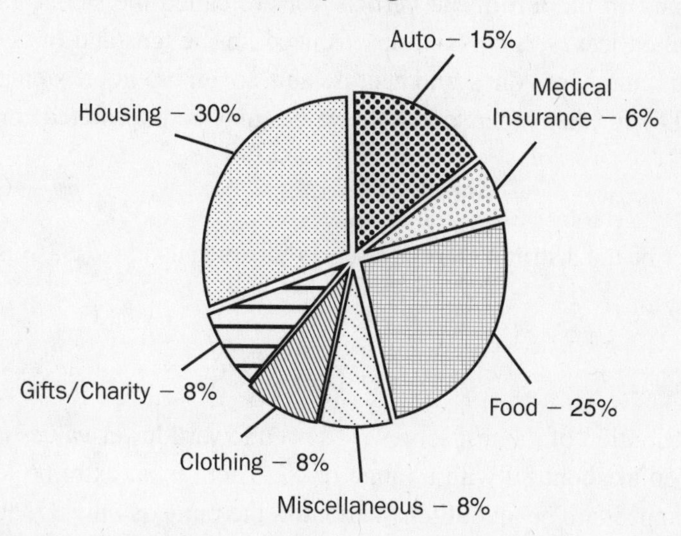

SOLUTION

To find the amount spent on housing, locate on the pie chart the percentage allotted to housing, or 30%. Then calculate 30% of $3,000 = $900. The family plans to spend $900 on housing.

STEMPLOTS

A **stemplot**, also called stem-and-leaf plot, can be used to display univariate data as well. It is good for small sets of data (about 50 or less) and forms a plot much like a histogram. The stemplot on the next page represents test scores for a class of 32 students.

Test Scores

3	3
4	
5	
6	3 7 9
7	2 2 5 7
8	1 2 6 8 8 8 9 9 9
9	0 0 0 1 3 3 4 5 5 6 7
10	0 0 0 0
Key: 6	3 represents a score of 63

The values on the left of the vertical bar are called the stems; those on the right are called leaves. Stems and leaves need not be tens and ones—they may be hundreds and tens, ones and tenths, and so on. A good stemplot always includes a key for the reader so that the values may be interpreted correctly.

PROBLEM

Describe the distribution of test scores for students in the class using the stemplot.

SOLUTION

The distribution of the test scores is skewed toward lower values (to the left). It is centered at about 89 with a range of 67. There is an extreme low value at 33, which appears to be an outlier. Without it, the range is only 37, about half as much. The test scores have a mean of approximately 85.4, a median of 89, and a mode of 100.

BIVARIATE DATA

Bivariate data examines the relationship between two variables. The two variables are called the **response variable** and the **explanatory variable**. The data is given as points (explanatory, response).

Response variable: measures the outcome of a study.

Explanatory variable: attempts to explain the response variable.

Examples:

How does studying affect the grade in a final exam? The explanatory variable is hours of study which explains the response variable, the grade. Someone who studies 8 hours and received a 94 would have the point (8, 94).

How does the average outside temperature affect your heating oil bill? The explanatory variable is average outside temperature in a month, which explains the response variable, the amount you spend for oil. A month whose average temperature was 34 degrees and in which the heating bill was $250 would have the point (34, 250).

PROBLEM

Charles works out on a treadmill that has many possible speeds. The treadmill also measures his pulse rate. Which is the explanatory variable and which is the response variable if his pulse rate was 125 when his speed was 7 mph? What would the data point be?

SOLUTION

The explanatory variable is speed and response variable is pulse rate. The point would be (7, 125).

A **scatterplot** shows the relationship between two quantitative variables measured on the same individuals.

The values of one variable appear on the horizontal (x) axis and the other variable appears on the vertical (y) axis. Each individual value appears as a point in the plot. The explanatory variable is placed on the x-axis and the response variable is placed on the y-axis.

Interpreting scatterplots

- Form – does the data appear linear or curved?
- Direction of association

Positive association–data goes up to the right

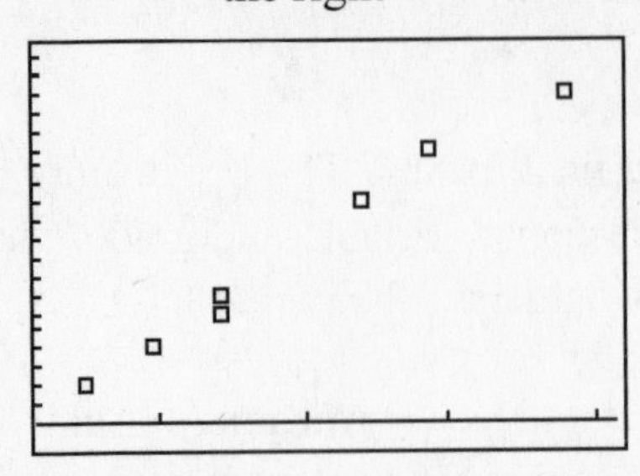

Negative association–data goes down to the right

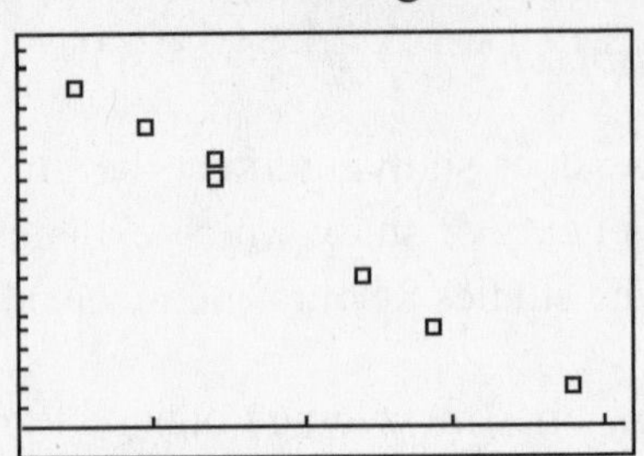

- Strength of an association – how closely the points follow a clear form. Both of the associations above are strongly linear.

Line of best fit (sometimes called the regression line):

The line of best fit is a straight line that describes how a response variable y changes as an explanatory variable x changes. Lines of best fit are used to predict the value of y for a given value of x. These lines require an explanatory variable and a response variable.

The closer the line of best fit comes to the data points, the stronger the association is:

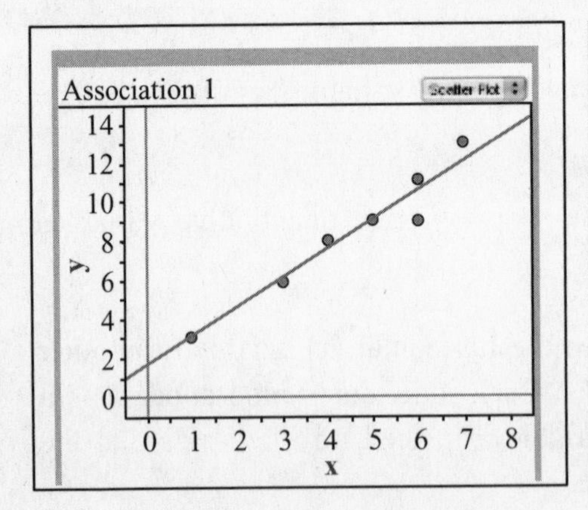

Strong positive association

Moderate negative association

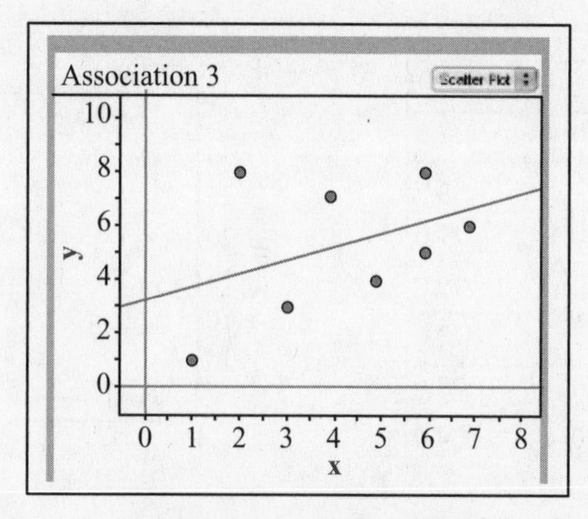

Weak positive association

- Meaning – an association between x and y means that as x changes, y changes. An association between x and y does not mean necessarily that x causes y.

Example:

There is a strong positive association between the number of firefighters sent to a fire and the amount of damage the fire does. However, sending firefighters to a fire does not cause damage.

The equation of the line of best fit describing data is in the familiar algebraic form $y = mx + b$ where x is the explanatory variable, y is the response variable, and m is the slope. Students are not responsible for calculating the line of best fit – it will be given. You simply need to interpret it.

Interpretation of slope m: For every unit increase in x (the explanatory variable), the y (the response variable) changes by m.

Example:

Over a 10-minute period in a busy bank, several tellers go to lunch and the line of customers becomes longer. The number of people in line at various times during that period is given in the table on the next page and the scatterplot of the data is shown.

People in line

	time	people
1	0.0	12.0
2	1.5	18.0
3	3.0	17.0
4	4.0	19.0
5	6.5	23.0
6	8.0	27.0
7	10.0	34.0

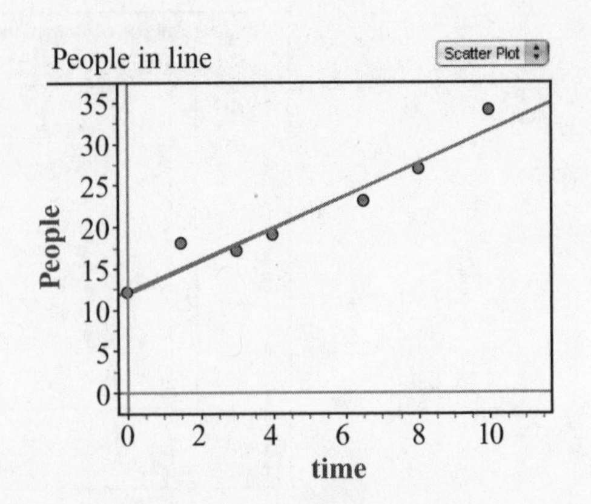

The association would be described as a strong positive association.

If the line of best fit equation is $y = 1.96x + 12.2$ where x represents the time and y represents the number of people in line, the slope of the line 1.96 is interpreted as: for every minute, the line increases by 1.96 people.

To predict the number of people in line at 7 minutes, we let $x = 7$ in the equation $y = 1.96x + 12.2$ and we get 25.92. So we predict that there were 26 people in line at 7 minutes. Since the association appears to be strong, we are fairly confident in our prediction. When we try to predict the value of y using a value of x between several given values of x, we call that **interpolation.**

To predict the number of people in line at 15 minutes, we let $x = 15$ in the equation $y = 1.96x + 12.2$ and we get 41.6. So we predict that there were 42 people in line at 15 minutes. However, even though the association is strong, we are less confident in that because nothing is known beyond the 10-minute mark. For instance, the tellers might come back from lunch and the line would be shorter. When we try to predict the value of y using a value of x outside the given values of x, we call that **extrapolation.**

PROBLEM

1. A study was done with 16 cars measuring their weight in pounds versus the mileage in miles per gallon. The scatterplot on the next page shows the line of best fit.

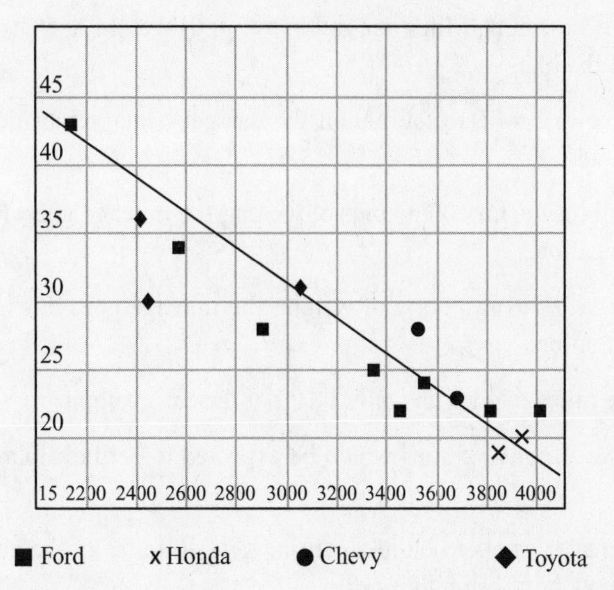

Which type of car has the greatest difference between its actual mileage and its predicted mileage?

(A) Honda

(B) Chevy

(C) Toyota

(D) Ford

2. Using the scatterplot above, if a new car was added to the study that weighed 3,750 pounds and got 40 miles to the gallon, how would the association be affected?

(A) becomes stronger with a steeper slope

(B) becomes weaker with a steeper slope

(C) becomes stronger with a more shallow slope

(D) becomes weaker with a more shallow slope

3. The line of best fit is: Miles per gallon = -0.01Weight + 59.86. Interpret the slope of the line.

   (A) For every mile to the gallon the car gets, the weight goes up by 100 pounds.

   (B) For every extra 100 pounds of weight, the mileage goes down by 1 mile per gallon.

   (C) For every extra pound of weight, the mileage goes down by 100 miles per gallon.

   (D) The mileage goes down by 1/100 of the car's weight.

4. A car that weighs 1.5 tons would be expected to get how many miles to the gallon?

5. If a carmaker wants to build a car that gets 40 miles to the gallon, approximately how heavy should it be?

SOLUTIONS

1. **(C)** The car whose data point is farthest from the line weighs a little more than 2,400 pounds and gets 30 mpg. It is (C) Toyota.

2. **(D)** The new point would be to the upper right of the graph. It would weaken the relationship and it would act as a magnet to pull the line towards it, making the slope more shallow.

3. **(D)** The slope of –0.01 has the interpretation that for every additional pound, the mileage goes down by 0.01 mpg. For every additional 100 pounds, the mileage goes down by 1 MPG.

4. Mileage = $-(3000) + 59.86 = 29.86$ MPG

5. $40 = -\,0.01x + 59.86 \Rightarrow 0.01x = 19.86 \Rightarrow x = \dfrac{19.86}{0.01} = 1{,}986$ lbs.

PROBLEM

Bears were anesthetized, tagged, and weighed. At different time intervals, the process was repeated to see the association between age in months and weight in pounds as shown in the scatterplot on the next page. The line of best fit is drawn in as well.

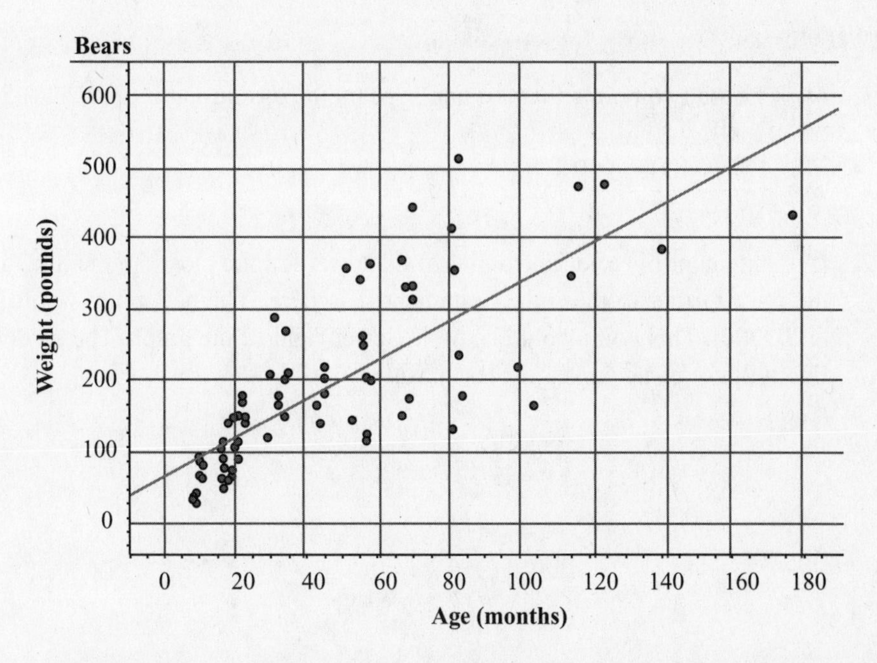

1. Which of the following values is closest to the monthly increase in weight of the bears?

   (A) 2.5

   (B) 5

   (C) 10

   (D) 20

2. Which of the following statements is true?

   I. The line of best fit is a better predictor of weights of young bears rather than older bears.

   II. If a new bear that is 15-years-old and weighs 200 pounds is added to the study, the direction of the association will become negative.

   (A) I only

   (B) II only

   (C) I and II

   (D) neither I nor II

SOLUTIONS

1. **(A)** Choose two points that the line appears to pass through: (120, 400) and (160, 500).

   $$m = \frac{500-400}{160-120} = \frac{100}{40} = 2.5.$$

2. **(A)** Statement I is true. The points at the lower left are closer to the line than the ones further to the right. Statement II is false. The new point would be (180, 200). This point would be to the lower right of the graph. The association will not be negative, just less positive.

Drill Questions

1. What is the difference between the mean and median of the following sample?

   {19, 15, 21, 24, 11, 18}

   (A) 0
   (B) 0.5
   (C) 1
   (D) 4

2. A car dealer asks its customers to rate their new car two months after they purchased it. They rate it on a 1—10 scale where 1 means very unhappy and 10 means very happy. The results are shown in the chart below. Arrange the mean, median, and mode from highest to lowest.

Rating	Number
10	2
9	12
8	17
7	14
6	21
5	5
4	1
3	0
2	1
1	2

(A) mean – median – mode
(B) mode – median – mean
(C) mean – mode – median
(D) median – mean – mode

3. Mrs. Smith teaches a class of 150 students. On a recent exam she administered to her class, 100 students scored 90 or better; 40 students scored between 80 and 89; and the remaining students scored between 70 and 79. Which of the following statements is correct concerning the students' exam scores?

   (A) The median equals the mean.
   (B) The mean is less than the mode.
   (C) The median is greater than the mean.
   (D) The median is greater than the mode.

4. What is the standard deviation of the data set: {–2, 0, 3, 5, 9}?

   (A) 3.54
   (B) 4.30
   (C) 12.5
   (D) 18.5

5. In a math class, the teacher curved an exam raising every student's score by 5 points. When the changed scores were compared to the original scores, which of the following would be 5 points higher?

   (A) Mean, but not standard deviation
   (B) Standard deviation, but not mean
   (C) Both mean and standard deviation
   (D) Neither mean nor standard deviation

6. A random sample of 25 people from the Sunnybrook Retirement Home were surveyed. The histogram on the next page displays the ages of these residents.

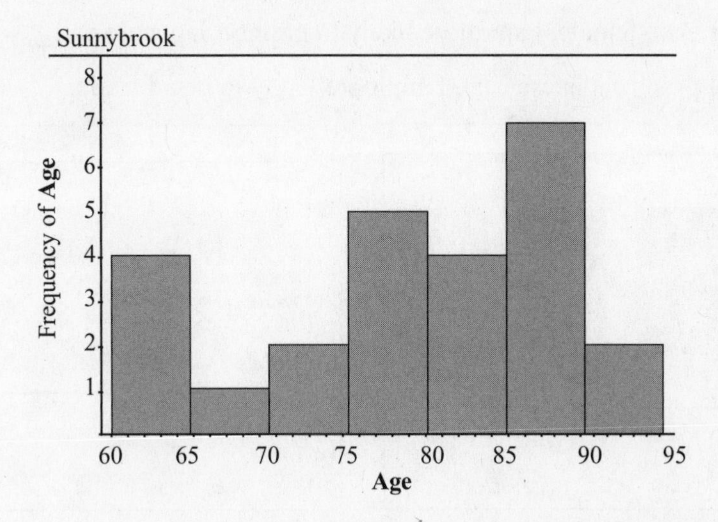

In which of the following intervals is the median age of these residents?

(A) 70 to less than 75
(B) 75 to less than 80
(C) 80 to less than 85
(D) 85 to less than 90

7. The number of years of experience in the sales department of a new car dealer is shown by the stemplot below (where 1 | 5 corresponds to 15 years). What is the difference between the mean and median number of years?

0	112356778
1	02359
2	135589

(A) 0
(B) 1
(C) 2
(D) 3

8. A study was done to compare the percentage of men and women who purchase American, European, and Japanese cars. The bar graph shown on the left on the next page shows the make of each car and the percentage of males and female who prefer it. The bar graph on the right shows the preference of men and women regarding the different makes. Which of the following statements is true?

I. If I am a female, I am more likely to prefer a Japanese car.

II. If I prefer Japanese cars, I am more likely to be a female.

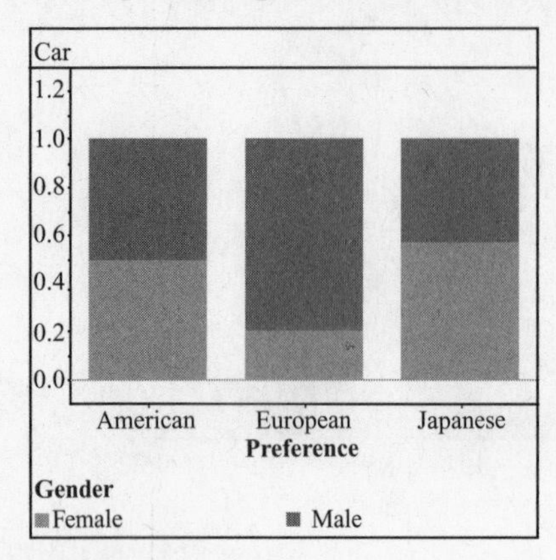

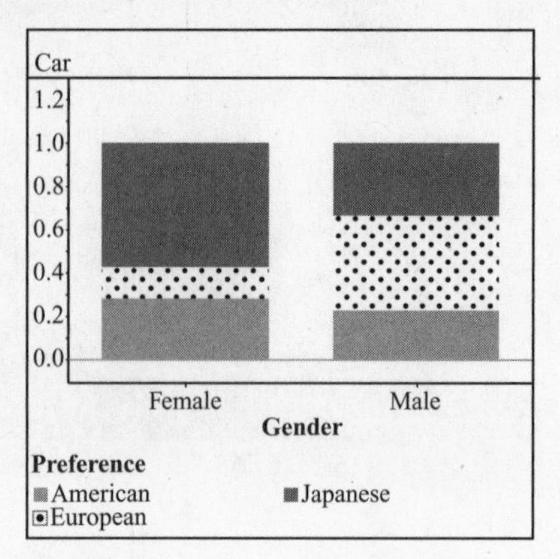

(A) I only
(B) II only
(C) I and II
(D) Neither

9. How many of the following sets of data would decidedly *not* graph a normal curve?

   I. Lengths of great white sharks

   II. The length of time it takes to do a transaction at a post office

   III. Body temperatures of people

   (A) 0
   (B) 1
   (C) 2
   (D) 3

10. The amount of time students in a college statistics class study for the final exam is approximately normally distributed with $\bar{x} = 8$ and $s = 2.5$ hours. Which represents the greatest percentage?

    (A) students studying less than 3 or more than 13 hours
    (B) students studying more than 8 hours
    (C) students studying less than 10.5 hours
    (D) students studying between 5.5 and 13 hours

11. In a final exam given to 400 students, the mean score was 75 with a standard deviation of 5. If the top grade of an A is defined as 85 or higher and an F is defined as 65 or lower, how many students get either an A or fail the exam?

(A) 5
(B) 10
(C) 20
(D) 40

12. Which scatterplot has the strongest positive linear association?

(A)
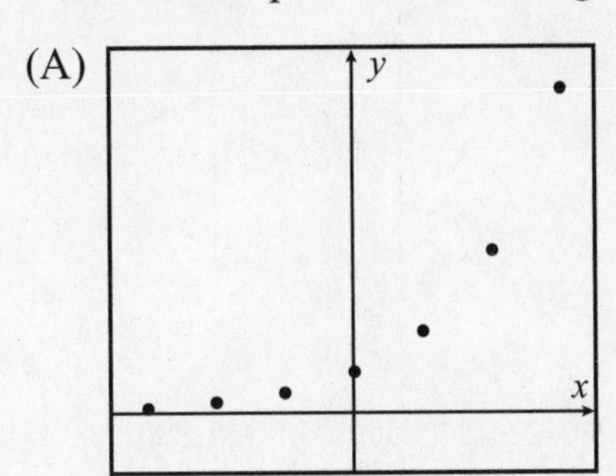

(C)
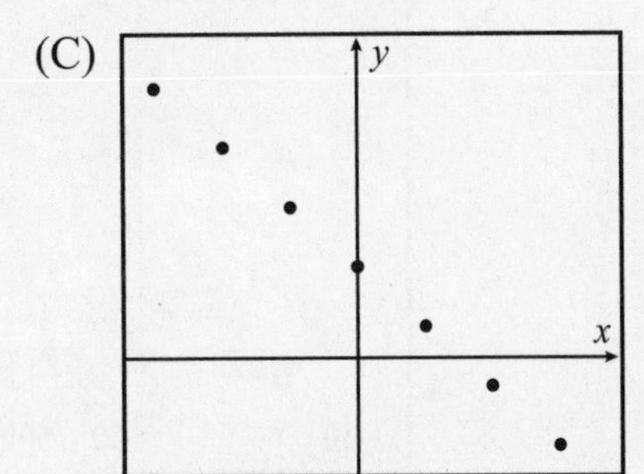

(B)
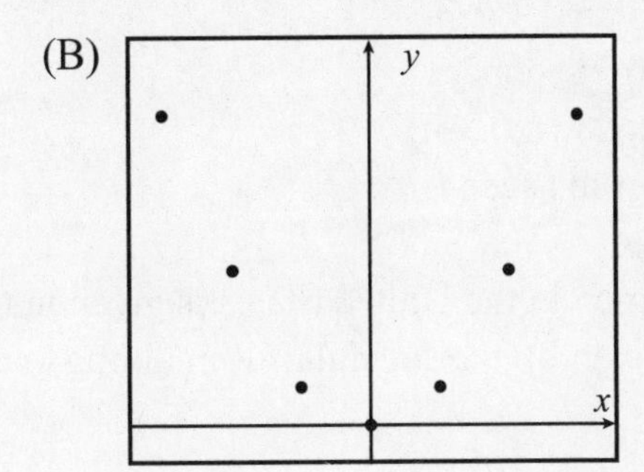

(D)
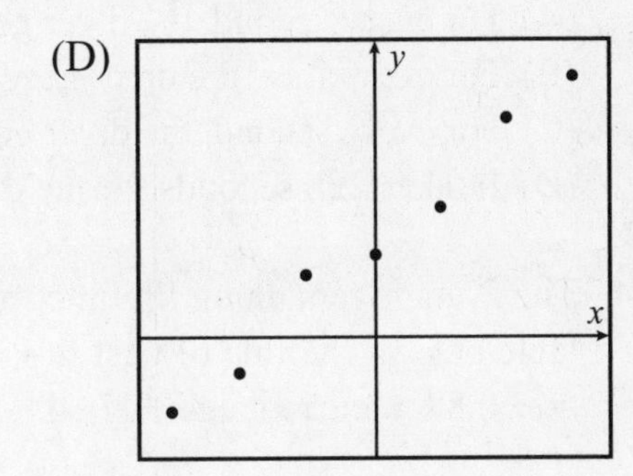

13. The depth of a diver ascending to the surface is shown in the scatterplot below. The line of best fit is Depth $= 228 - 0.3t$ where t is measured in seconds and depth is measured in feet. Interpret the slope of the line.

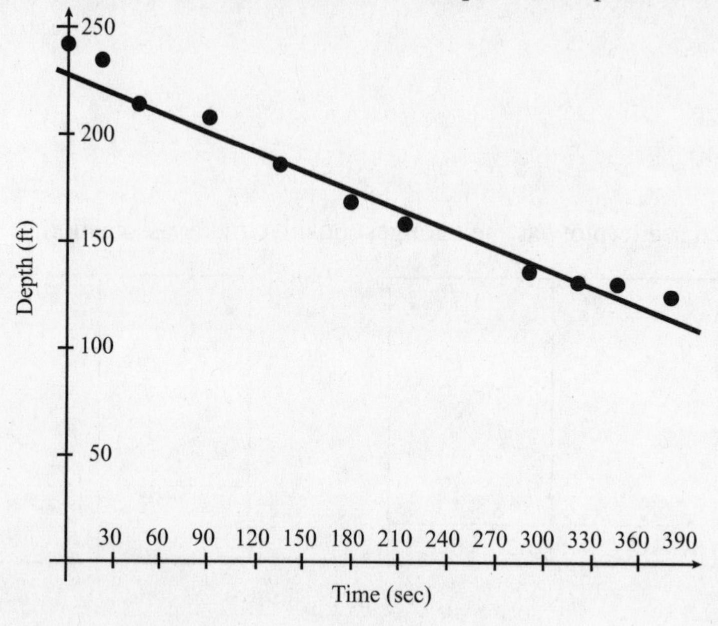

(A) For every second, the diver goes up 0.3 ft.
(B) For every foot, the diver needs 0.3 seconds.
(C) For every second, the diver goes down 0.3 ft.
(D) It takes 228 seconds for the diver to ascend.

14. The average remaining lifetime for men in the United States is given in the table below. The line of best fit is given by the formula: Remaining years $= -0.88 \times$ current age $+ 73.4$.

Current age	Remaining years
0	74.9
15	60.6
35	42.1
50	26.3
65	16.8
74	10.2

Using this formula, which person is expected to live the longest?

(A) a 5-year-old
(B) a 25-year-old
(C) a 45-year-old
(D) a 70-year-old

15. Body Mass Index (BMI) is a calculation that relates a person's height and weight. The table below shows the percent of U.S. males age 18 years or older who were considered obese in the years indicated, judged on the basis of BMI. The line of best fit is given by the formula:

Percent obese $= 0.775 \cdot \text{year} - 1529.3$.

Year	Percent obese
2003	22.7
2004	23.9
2005	24.9
2006	25.3
2007	26.5
2008	26.6
2009	27.6

In what year would we expect one-third of the U.S. male population age 18 or older to be considered obese?

(A) 2014
(B) 2015
(C) 2016
(D) 2017

Answers to Drill Questions

1. **(B)** $\bar{x} = \frac{19+15+21+24+11+18}{6} = 18$

   Sorted data: {11, 15, 18, 19, 21, 24} Median is average of 18 and 19 which is 18.5. Difference: 18.5 − 18 = 0.5

2. **(D)** $\bar{x} = \frac{2(10)+12(9)+17(8)+14(7)+21(6)+5(5)+1(4)+1(2)+2(1)}{6}$

   $= \frac{521}{75} = 6.95$

   Median is the data #38 $=7$
   Mode = 6

3. **(C)** There were only 150 − 100 − 40 = 10 students who scored 70 − 79. We know nothing about the mode. It could be that all 40 people scored an 80. But the distribution is skewed to the left which shows its mean is less than the median.

4. **(B)**

   $\bar{x} = \frac{-2+0+3+5+9}{5} = \frac{15}{5} = 3$

   $s^2 = \frac{(-2-3)^2+(0-3)^2+(3-3)^2+(5-3)^2+(9-3)^2}{4}$

   $= \frac{25+9+0+4+36}{4} = \frac{74}{4} = 18.5$

   $s = \sqrt{18.5} = 4.30$

5. **(A)** If there are n students, the sum of their scores will be $5n$ higher and the average is $\frac{5n}{n} = 5$ points higher. The standard deviation is the spread of the data which has not changed.

6. **(C)** The median is in the 13th piece of data which occurs in the 80–85 grouping.

7. **(C)** The median is the average of the 10th and 11th pieces of data, 10 and 12, which is 11.

$$\bar{x}=\frac{2(1)+2+3+5+6+2(7)+8+10+12+13+15+19+21+23+2(25)+28+29}{20}$$

$$=\frac{260}{20}=13$$

Difference: 13 – 11 = 2.

8. **(C)** I. Right bar-graph, look at the female bar. About 60% of the females prefer Japanese cars.
   II. Left bar-graph, look at the Japanese bar. About 55% of the people preferring Japanese cars are female

9. **(B)** I. Some sharks will be short, most will be medium length, and some are long. It should appear normal.
   II. Most transitions take little time. There are fewer that will take longer. The data is skewed right.
   III. Most people will be 98.6 with a small standard deviation. It should appear somewhat normal.

10. **(C)** A. 5% B. 50% C. 84% D.81.5%

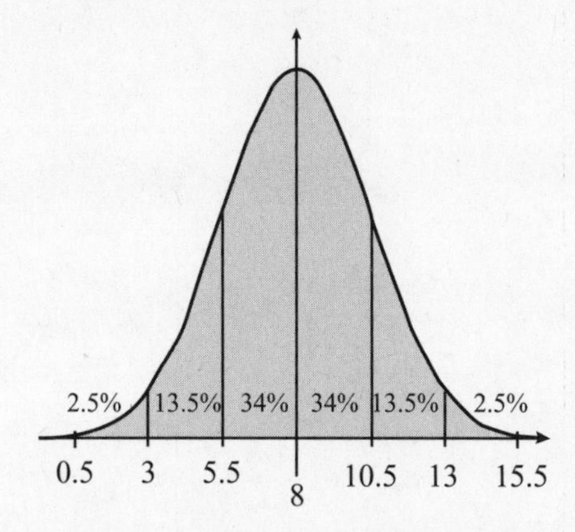

11. **(C)** 85 is 2 standard deviations above the mean so 0.025(400), 10 students get A's. 65 is 2 standard deviations above the mean so 0.025(400), 10 students get F's. 20 students either get an A or an F.
12. **(D)** (A) and (B) are strong relationships, but not linear. (C) is a strong linear association, but negative.

13. **(A)** The definition of slope m is every unit change in t, the response variable change by m. Since m is negative, the depth is getting smaller, meaning the diver is going up.

14. **(D)**
(A) Remaining years = $-0.88(5) + 73.4 + 5 = 74$
(B) Remaining years = $-0.88(25) + 73.4 + 25 = 76.4$
(C) Remaining years = $-0.88(45) + 73.4 + 45 = 78.8$
(D) Remaining years = $-0.88(70) + 73.4 + 70 = 81.7$

15. **(C)** Percent obese = 0.775 year – 1529.3 = 33.333
0.775 year = 1562.633
Year = 2016

CHAPTER 6

Financial Mathematics

CHAPTER 6

FINANCIAL MATHEMATICS

Percentages provide the basis for most problems involving financial situations. Percentages can be expressed as a percent, a fraction (usually over 100), or a decimal. Remember to include the leading zero for the decimal equivalent of percentages less than 10% (i.e., 5% = 0.05, not 0.5).

RATES

The word *rate* is used frequently in our everyday lives. Examples are *tax rates, hourly rates, inflation rates, vehicle accident rate,* and *consumption rate.* Rates are written as fractions.

Example:

A hybrid car travels 250 miles on 6 dollars of gas. The miles-to-dollars rate for this car is $\frac{250 \text{ miles}}{\$6}$.

An SUV travels 350 miles on 20 dollars of gas. The miles-to-dollars rate for this car is $\frac{350 \text{ miles}}{\$20}$.

A unit rate is a rate in which the denominator is 1. Unit rates make comparisons easier.

Example:

The hybrid car above gets $\frac{250}{6} = 41.67$ miles on \$1 of gasoline while the SUV gets $\frac{350}{20} = 17.5$ miles on \$1 of gasoline.

PROBLEM

A doctor's assistant earns $1,350 for working a 45-hour week. What is the assistant's hourly rate of pay?

SOLUTION

$$\frac{\$1,350}{45 \text{ hr}} = \$30/\text{hr}.$$

PROBLEM

A contractor is installing a bathroom and estimates that there will be 40 hours of labor to complete the project. He estimates the cost of the job at $13,250. When the work is actually done, there are 45 hours of labor. How much should he charge?

SOLUTION

The unit rate is $\frac{\$13,250}{40 \text{ hr}} = \$331.25/\text{hour.}$ = $331.25/hour. For 45 hours, the charge will be 45(331.25) = $14,906.25.

The unit price of a product is its cost per unit of measure. To find the unit price, write the rate as a unit rate. 5 pounds of sugar costing $1.45 has a unit rate of $\frac{\$1.45}{5 \text{ pounds}} = \$0.29/\text{pound}$. Consumers use unit pricing to determine the better buy. The more economical purchase is the product with the lower unit price, typically associated with a larger size or quantity. However, products with smaller unit prices can also cost more in total, depending upon the size offered. For example, a perishable product becomes a better deal only if you need all of it. A fifty-pack of batteries may cost much less per unit, but will cost much more in the long run if you seldom need batteries and most of them pass their sell-by date unused.

PROBLEM

Which is the more economical purchase: a 20-ounce bottle of ketchup at \$1.49 or a 28-ounce bottle of ketchup at \$1.89?

SOLUTION

$$\frac{\$1.49}{20 \text{ ounces}} = \$0.0745/\text{ounce.} \qquad \frac{\$1.89}{28 \text{ ounces}} = \$0.0675/\text{ounce.}$$

The \$1.89 bottle of ketchup is the better deal.

PERCENTS

Problems involving percentages are best solved using basic equations. The word *is* translates to *equals*. The word *percent* is expressed as a fraction over 100, and the word *of* is expressed as multiplication. The unknown is expressed as x. 10% can be expressed as $\frac{10}{100}$ or 0.10, while 200% can be expressed as $\frac{200}{100}$ or 2.

PROBLEMS

1. What is 25% of 12?
2. 14 is 20% of what number?
3. 6 is what percent of 20?

SOLUTIONS

1. $\frac{25}{100} \cdot 12 = x \Rightarrow x = \frac{300}{100} = 3$
2. $14 = \frac{20}{100} \cdot x \Rightarrow 1400 = 20x \Rightarrow x = 70$
3. $6 = \frac{x}{100} \cdot 20 \Rightarrow 20x = 600 \Rightarrow x = 30$

There are many financial situations that involve increasing or decreasing an amount by a certain percentage.

% Increase - Find the difference between the new price and the original price and then determine what percent this is of the original price.

$$\% \text{ increase} = \frac{\text{New price} - \text{Original price}}{\text{Original price}}$$

% Decrease – Find the difference between the original price and new price and then determine what percent this is of the original price.

$$\% \text{ decrease} = \frac{\text{Original price} - \text{New price}}{\text{Original price}}$$

PROBLEM

A house originally cost $40,000. The house sold for $150,000. By what percent did the house increase in value?

SOLUTION

$\% \text{ increase} = \frac{150{,}000 - 40{,}000}{40{,}000} = 2.75$. So the house increased by 275%.

PROBLEM

A new car costs $24,000. As a trade-in, the car would only be worth $15,000. By what percent did the car decrease in value?

SOLUTION

$$\% \text{ decrease} = \frac{24{,}000 - 15{,}000}{24{,}000} = \frac{9{,}000}{24{,}000} = 0.375$$

The car decreased in value by 37.5%.

PROBLEM

The cost of Internet service was $45 a month last year. This year, it is $50. What is the percent increase in price?

SOLUTION

$$\% \text{ increase} = \frac{50 - 45}{45} = \frac{5}{45} = 0.111.$$

There is an 11.1% increase in price.

PROBLEM

Your dream cellphone is on sale for 20% off the original price. If the original price is $280, what is the sale price?

SOLUTION

$$\% \text{ decrease} = \frac{20}{100} = \frac{280 - N}{280}$$

$$28{,}000 - 100N = 5{,}600 \Rightarrow 100N = 22{,}400 \Rightarrow N = \$224.$$

PROBLEM

A shirt is on sale for $30. The advertisement says that this price represents a 25% decrease in price. What was the shirt's original price?

SOLUTION

$$\% \text{ decrease} = \frac{25}{100} = \frac{x - 30}{x} \Rightarrow 100x - 3{,}000 = 25x \Rightarrow 75x = 3{,}000 \Rightarrow x = \$40.$$

TAXES

We all have to pay taxes. While few people relish paying taxes, everyone enjoys having good schools, roads, and services. In the end, taxes improve our overall quality of life. There are many types of taxes and usually they are computed as a percentage of some quantity. Here are some of the common ones:

- Sales tax: a tax on goods and services. Sales tax is determined by the state and can vary from state to state. As of this writing, the tax rate varies from 0% (Alaska, Delaware, Montana, and Oregon) to 7.5% (California).
- Property tax: paid by anyone who owns land.
- Capital gains taxes: paid by people who have investments that have appreciated (gained value over time) at the time they are sold. Examples include stocks, bonds, and real estate. You pay capital gains taxes on the difference between the purchase price and the sale price. If you sell the investment for less than you paid, you have a capital gains loss.

- Income tax: paid by people who earn income. Income tax is usually not paid by people who earn less than a certain income or who have special situations such as a disability. The amount of money that a person makes is called the **gross income**. When we deduct the income tax and other taxes, we are left with the **net income.** Everyone pays federal income tax. Depending on where you live, you may also pay state taxes and local taxes.

In 2015, the federal tax rates for individual taxpayers are as follows:

If taxable income is between	The tax due is
0 – $9,225	10% of the taxable income
$9,226 – $37,450	$923 + 15% of the amount over $9,225
$37,451 – $90,750	$5,156 + 25% of the amount over $37,450
$90,751 – $189,300	$18,481 + 28% of the amount over $90,750
$189,301 – $411,500	$46,075 + 33% of the amount over $189,300
$411,501 – $413,200	$119,401 + 35% of the amount over $411,500
$413,201 +	$119,996 + 39.6% of the amount over $413,200

Example:

Carrie is single and has a gross income (all taxable) of $50,000. Based on the federal tax rate table above, she is in the category of someone making between $37,451 and $90,750. She is in the 25% tax bracket (paying $5,156 + 25% of the amount over $37,450).

To determine her income tax, we first find the amount of her income that is over $37,450.

$50,000 – $37,450 = $12,550

Carrie pays 5,156 + 0.25(12,550) = 5,156 + 3,137.50 = $8,293.50.

She ends up paying $\frac{8,293.50}{50,000} = 16.6\%$ of her salary in taxes.

PROBLEM

Wayne lives in Doylestown, Pennsylvania. Pennsylvania has a 3.07% personal income tax. Doylestown has a 1.025% local tax. If Wayne makes $110,000 a year, what is his net salary and what percent of his salary does he pay in taxes?

SOLUTION

Federal tax bracket: 28% $\quad 110{,}000 - 90{,}750 = 19{,}250$

Pay: $18{,}481 + 028(19{,}250) = 18{,}481 + 5{,}390 = \$23{,}871$

State Tax: $0.0307(110{,}000) = \$3{,}377$

Local Tax: $0.01025(110{,}000) = \$1{,}127.50$

Total Taxes: $23{,}871 + 3{,}377 + 1{,}127.50 = \$28{,}375.50$

$$\text{Percentage taxes} = \frac{\$28{,}375.50}{110{,}000} = 25.8\%$$

PROBLEM

Hans owns his own business and makes $175,000 a year in taxable income. He is considering taking on a new contract that would pay him an additional $50,000. What is the difference in federal taxes that he would pay with the additional income? What percentage of the additional $50,000 would be going to taxes?

SOLUTION

Income $= \$175{,}000$: Tax $= 18{,}481 + 028(175{,}000 - 90{,}750) = 18{,}481 + 0.28(84{,}250) = 18{,}481 + 23{,}590 = \$42{,}071$

Income $= \$225{,}000$: Tax $= 46{,}075 + 0.33(225{,}000 - 189{,}300) = 46{,}075 + 0.33(35{,}700) = 46{,}075 + 11{,}781 = \$57{,}856$

Difference: $57{,}856 - 42{,}071 = \$15{,}785$

$$\text{Percent taxes: } \frac{15{,}785}{50{,}000} = 31.6\%$$

MARKUPS AND MARKDOWNS

A **markup** is the difference in the amount the seller buys the product for and the amount at which the seller sells the product.

Markup = Selling price − Cost and $\% \text{ Markup} = \frac{\text{Markup}}{\text{Cost}}$.

All businesspeople markup any merchandise they sell. (If they didn't, they wouldn't make money.) If a store owner bought a candy bar for 50 cents and sold it for a dollar, his markup would be 1.00 − 0.50 = 0.50. His percent markup would be $\frac{\text{Markup}}{\text{Cost}} = \frac{0.50}{0.50} = 100\%$.

A **markdown** is a reduction from the original selling price. There are many reasons that a retailer would mark down a product. For a car dealer, it could be the change in model year, and the dealer may need to sell all of last year's models to make room for the new year's models. In a supermarket, fruit is perishable. The retailer can either sell the fruit that's nearing its shelf-life limitations at a discount or throw it out. Many businesses will lure a customer in with a lower price in the hope that the customer will purchase other items at full price.

A markdown (or discount) is the amount the seller will take off the price of the product to give an incentive for a buyer to purchase it.

Markdown = Original price − New price.

The percentage markdown is $\frac{\text{Markdown}}{\text{Original Price}}$.

PROBLEM

A dealer sells a refrigerator at its list price of $1,200. The dealer's cost was $850. What is the markup and percentage markup?

SOLUTION

The markup is 1,200 − 850 = $350. His percentage markup is given by $\frac{\text{Markup}}{\text{Cost}} = \frac{350}{850} = 41.2\%$.

PROBLEM

A new car dealer purchases a car for $25,000. The sticker price on the car (what the dealer tries to sell the car for) is $30,000. The dealer discounts the car by 5%. What is his percentage markup on the car?

SOLUTION

x is the discounted price

$$\frac{5}{100} = \frac{30{,}000 - x}{30{,}000} \Rightarrow 3{,}000{,}000 - 100x = 150{,}000 \Rightarrow 100x = 2{,}850{,}000 \Rightarrow$$

$$x = \$28{,}500$$

Markup = $28,500 − $25,000 = $3,500 so percentage markup =

$$\frac{\text{Markup}}{\text{Cost}} = \frac{3{,}500}{25{,}000} = 14\%.$$

PROBLEM

A street vendor sells flowers for $10 a bunch. At the end of the day, he sells the flowers for $4 a bunch. What is the markdown and percentage markdown?

SOLUTION

The markdown is $10 − $4 = $6. The percentage markdown is

$$\frac{\text{Markdown}}{\text{Original Price}} = \frac{6}{10} = 60\%$$

When you operate a business or run a home, you either have a profit, a loss, or you break even.

If your sales (the money you take in) are greater than your expenses (the money you have to pay), you have a profit. If your expenses are greater than your sales, you have a loss. If your sale equals your expenses, you break even.

Suppose a car dealership has $100,000 in expenses per month. If the markup on a car is approximately $2,000, they would need to sell 50 cars a month to break even. If the dealership's goal is to have a $20,000 profit for the month, they would need to sell the 50 cars to break even and then another 10 cars for the $20,000 profit. So they would need to sell 60 cars for the month.

PROBLEM

It costs Mr. Meltzer $800 a week to run his household (rent, food, electricity, etc.). His salary is $40,000 annually. What percent raise will get him to the break-even point?

SOLUTION

$\frac{40,000}{52} = 769.23$ a week which is a loss of $30.77

$x\%$ of 769.23 is $30.77 \Rightarrow \frac{x}{100} - 769.23 = 30.77 \Rightarrow x = \frac{3,077}{769.23} = 4.$

If he gets a 4% raise, he will break even.

TYPES OF INTEREST

When you deposit money into a bank, you are allowing the bank to use your money. The bank may in turn lend that money to other customers to purchase a house or a car. The bank pays you for using your money. The amount that the bank pays you is called **interest**. If you use the bank for savings, the bank pays you interest. If you are the one who gets a loan from the bank, you pay the bank interest for the use of their money.

The amount you borrow from a bank or the money you deposit in a bank is called the **principal**. The amount of interest paid is usually a percent of the principal.

Simple Interest

The simple interest formula is $I = P \times r \times t$ where P is the principal, r is the interest rate, and t is the time period.

The interest rate is usually given per year, called the **annual percentage rate** (APR). When r is given as a percentage per year, the time is also expressed in years. It is possible for the rate to be given in other time periods, such as months. In that case, time would have to be measured in months. Bank interest rates are usually given as an APR unless otherwise stated.

PROBLEM

Calculate the simple interest earned in two years on a deposit of $500 if the APR is 3%.

SOLUTION

$P = 500, r = 3\% = 0.03, t = 2$

$I = P \cdot r \cdot t = 500(0.03)(2) = 30$. So the simple interest is $30.

PROBLEM

Calculate the simple interest due on a 6-month loan of $2,500 if the APR is 4.5%.

SOLUTION

$$P = 2{,}500,\ r = 4.5\% = 0.045,\ t = \frac{6 \text{ months}}{1 \text{ year}} = \frac{6 \text{ months}}{12 \text{ months}} = \frac{1}{2}$$

$I = 2{,}500(0.045)(0.5) = 56.25$. So the simple interest is $56.25.

(Note that rather than work in years, we could work in months and still arrive at the same answer.)

$$P = 2{,}500,\ r = \frac{4.5\%}{\text{year}} = \frac{0.045}{12 \text{ months}},\ t = 6 \text{ months}$$

$$I = 2{,}500\left(\frac{0.045}{12}\right)(6) = \$56.25.$$

PROBLEM

If I earned \$250 simple interest on a 3-year deposit of \$5,000, what was the interest rate?

SOLUTION

$P = 5{,}000,\ I = 250,\ t = 3$

$$250 = 5{,}000(r)(3) \Rightarrow r = \frac{250}{15{,}000} = 0.0167 = 1.67\%$$

If the time period of a loan with an annual interest rate is given in days, it is necessary to convert the time period of the loan to a fractional part of a year. Usually, to simplify calculations, the value of t in years is divided by 360, an approximation to the number of days in the year (12 months times an average of 30 days per month).

PROBLEM

Calculate the simple interest on a 45-day loan of \$4,000 if the interest rate is 7%.

SOLUTION:

$$P = 4{,}000,\ r = 7\% = 0.07,\ t = \frac{45 \text{ days}}{360 \text{ days}} = \frac{45}{360}$$

$I = 4{,}000(0.07)\left(\frac{45}{360}\right) = 35$. So the simple interest due is \$35.

Compound Interest

Simple interest is given on the original principal only. Once interest is applied to the principal, however, the principal changes. It is only fair to give interest to the new principal. The process of applying interest to the original principal and to any interest that has occurred is called *compounding* and the interest is called **compound interest**.

PROBLEM

You deposit $1,000 in a savings account that earns 4% interest, compounded annually, for 4 years. How much will you have at the end of the 4 years?

SOLUTION

During the first year, the interest is given by $I = P \times r \times t = 1{,}000(0.04)(1) = \$40$. So at the end of the first year, the principal is now $1,040.

During the second year, the interest is given by $I = P \times r \times t = 1{,}040(0.04)(1) = \$41.60$. Instead of $40 interest, we have made $41.60 because our interest is based on a principal of $1,040 rather than $1,000. So at the end of the second year, the principal is now $1,081.60.

During the third year, the interest is given by $I = P \times r \times t = 1{,}081.60(0.04)(1) = \$43.26$. Our interest has gone up again and the principal is now $1,124.86.

Finally during the fourth year, the interest is given by $I = P \times r \times t = 1{,}124.86(0.04)(1) = \$44.99$. Our interest has gone up again and the maturity value is $1,169.85.

Using simple interest, we would only have received $40 interest for each of the 4 years and our maturity value would have been $1,160. Compounding gave us an extra $9.85. It's not much, but when interest is compounded for many years, when the interest rate is high, or the principal is high, compounding yields much greater amounts than simple interest does.

Compounding does not always occur annually. Interest can be given more frequently during the year. The frequency with which the interest is compounded is called the **compounding period**.

PROBLEM

Again you deposit \$1,000 in a bank at 4% APR, but now the interest is **compounded quarterly**. This means that at the end of each quarter of a year (3 months), interest will be applied. Our interest will be given 4 times a year meaning that each quarter we will get $\frac{4\%}{4} = 1\%$. How much will we have at the end of one year?

SOLUTION

End of first quarter: $I = P \times r \times t = 1{,}000(0.01)(1) = \$10$

$$A = P + I = 1{,}000 + 10 = \$1{,}010$$

End of second quarter: $I = P \times r \times t = 1{,}010(0.01)(1) = \$10.10$

$$A = P + I = 1{,}010 + 10.10 = \$1{,}020.10$$

End of third quarter: $I = P \times r \times t = 1{,}020.10(0.01)(1) = \$10.20$

$$A = P + I = 1{,}020.10 + 10.20 = \$1{,}030.30$$

End of fourth quarter: $I = P \times r \times t = 1{,}030.30(0.01)(1) = \$10.30.$

$$A = P + I = 1{,}030.30 + 10.30 = \$1{,}040.60$$

Compared to the \$1,040 amount when compounded annually, we have made an extra 60 cents when compounding quarterly. Compounding using even shorter periods of time can give savers more money, as we will soon see.

Compound Amount Formula

The compound amount formula is given by $A = P\left(1 + \frac{r}{n}\right)^{nt}$ where A is the compound amount, P is the principal, r is the APR, n is the number of compounding periods per year, and t is the number of years.

PROBLEM

Using the compounding periods listed below, find the future value of a $1,000 deposit invested for 10 years in a savings account earning 4% interest.

(A) Simple

(B) Annually

(C) Semiannually

(D) Quarterly

(E) Monthly

(F) Daily

SOLUTION

Interest compounded	n	Formula	A
(A) Simple – no compounding		$A=1{,}000[1+0.04(10)]$	$1, 400.00
(B) Annually	1	$A=1{,}000\left(1+\frac{0.04}{1}\right)^{1(10)}$	$1, 480.24
(C) Semiannually	2	$A=1{,}000\left(1+\frac{0.04}{2}\right)^{2(10)}$	$1, 485.95
(D) Quarterly	4	$A=1{,}000\left(1+\frac{0.04}{4}\right)^{4(10)}$	$1, 488.86
(E) Monthly	12	$A=1{,}000\left(1+\frac{0.04}{12}\right)^{12(10)}$	$1, 490.83
(F) Daily	360	$A=1{,}000\left(1+\frac{0.04}{360}\right)^{360(10)}$	$1, 491.79

Note that the greater number of compound periods, the larger the interest amount becomes. No banks use simple interest and you cannot get a loan using simple interest. There is only about a $12 difference between the value of the principal when compounded annually and compounded daily. Still, when the principal and number of years are large, there can be quite a big difference in the future value using the different methods. We also see that as we increase the number of compounding periods, the larger the compound amount becomes. But it gets larger by a smaller amount. There is less than a one dollar difference between the amounts you receive by compounding daily as opposed to compounding monthly. Over 10 years, that amount is negligible. It is for that reason that most banks today compound daily.

FUTURE VALUE AND MATURITY VALUE

When you borrow money, the total amount to be repaid to the lender is the sum of the principal and interest. When you withdraw money from a bank, you will receive both the principal and the interest. We call this sum the **future value** or **maturity value**.

The future or maturity value for simple interest is given by $A = P + I$ where A is the amount after the interest, and I has been added to the principal, P. The formula can also be expressed as $A = P + P \times r \times t$ or $A = P(1 + rt)$.

PROBLEM

Calculate the maturity value of a simple interest, 15-month loan of $7,500 if the interest rate is 6.5%.

SOLUTION

$$P = 7{,}500,\ r = 0.065,\ t = 15 \text{ months} \cdot \frac{1 \text{ year}}{12 \text{ months}} = \frac{15}{12} \text{ years}$$

$$I = 7{,}500\,(0.065)\left(\frac{15}{12}\right) = \$609.38$$

A = 7,500 + 609.38 = 8,109.38. The borrower needs to pay back $8,109.38.

PROBLEM

At the birth of their child, Greg, the Goldbergs place $5,000 into a simple interest account that earns 3.5% APR. When Greg graduates from high school at age 18, they plan to give him the money in the account. What is the future value of the deposit?

SOLUTION

$P = 5{,}000,\ r = 0.035,\ t = 18$

$I = P \cdot r \cdot t = 5{,}000(0.035)(18) = \$3{,}105$

$A = P + I = 5{,}000 + 3{,}105 = 8{,}105$. Greg will receive \$8,105.

PROBLEM

The maturity value of a 3-month loan of \$8,000 is \$8,170. What is the simple interest rate?

SOLUTION

$$A = 8{,}170,\ P = 8{,}000,\ t = 3 \text{ months} \cdot \frac{1 \text{ year}}{12 \text{ months}} = \frac{3}{12} = \frac{1}{4} \text{year}$$

$$A = P + I \Rightarrow 8{,}170 = 8{,}000 + I \Rightarrow I = 170$$

$$I = P \cdot r \cdot t\ 170 \Rightarrow 8{,}000\left(\frac{1}{4}\right)r$$

$$r = \frac{170}{2{,}000} = 0.085.$$ The interest rate is 8.5%.

PROBLEM

Your property tax bill is \$3,200. The county charges a penalty of 16% simple interest for late payments. How much would you pay if you pay the bill one month late?

SOLUTION

$$P = 3{,}200,\ r = 0.16,\ t = 1 \text{ month} \cdot \frac{1 \text{ year}}{12 \text{ months}} = \frac{1}{12} \text{ years}$$

$$A = P(1 + rt) = 3{,}200\left[1 + 0.16\left(\frac{1}{12}\right)\right] = \$3{,}242.67.$$

PROBLEM

At the birth of their child, Greg, the Goldbergs place $5,000 into a compound interest account that earns 3.5% APR. When Greg graduates from high school at age 18, they plan to give him the money in the account. What is the difference in the future value of the deposit if it is compounded annually and compounded daily?

SOLUTION

$$\text{Annually: } A = 5{,}000\left(1+\frac{0.035}{1}\right)^{1(18)} = \$9{,}287.45$$

$$\text{Daily: } A = 5{,}000\left(1+\frac{0.035}{360}\right)^{360(18)} = \$9{,}387.77$$

Difference: $100.32

PROBLEM

Ted, who is 25 years old, places $2,500 into a retirement account that he will redeem at age 65. He has a choice of 6.5% APR compounded quarterly or 6.4% APR compounded daily. Which one will give him the larger future value?

SOLUTION

$$\text{Quarterly: } A = 2{,}500\left(1+\frac{0.065}{4}\right)^{4(40)} = \$32{,}963.20$$

$$\text{Daily: } A = 2{,}500\left(1+\frac{0.064}{360}\right)^{360(40)} = \$32{,}332.18$$

The 6.5% APR compounded quarterly will give him the larger future value.

PROBLEM

Ted purchases a $40,000 car. He puts a $10,000 down payment on the car and finances the remaining balance. What is the future value of the loan if he finances the car at 5.2% interest for 4 years compounded daily?

SOLUTION

$$A = 30,000\left(1 + .\frac{0.052}{360}\right)^{360(4)} = 36,935.84$$

The value of the loan is $36,935.84 if he pays it back at the end of the loan period.

Since he makes monthly payments, he will actually pay back less money.

Some banking institutions advertise **continuous compounding,** which means the number of compounding periods per year gets very, very large. When compounding continuously, rather than use the formula $A = P\left(1+\frac{r}{n}\right)^{nt}$, we use the formula $A = Pe^{rt}$ where e is approximately equal to 2.71828. Unless the principal is large and/or the length of time is large, continuous compounding yields little more than compounding daily does.

PROBLEM

Find the difference in the future value of a $10,000 investment for 30 years at 5% APR compounded daily and compounded continuously.

SOLUTION

Daily: $A = 10,000\left(1 + \frac{0.05}{360}\right)^{360(30)} = \$44,812.22$

Continuously: $A = 10,000(2.71828)^{0.05(30)} = \$44,816.85$

Difference: \$4.63

When interest is compounded, the annual rate of interest (APR) is sometimes called the **nominal rate.** The **effective rate** is the simple interest rate that would yield the same amount of interest after one year. When a bank advertises a 3% annual interest rate compounded daily, yielding 3.05%, the nominal interest rate is 3% and the effective rate is 3.05%.

PROBLEM

Find the effective interest rate on an investment compounded quarterly with an APR of 4.25%.

SOLUTION

The principal amount makes no difference. Use \$100 to make the numbers easy.

$$A = 100\left(1 + \frac{0.0425}{4}\right)^{4(1)} = \$104.32$$

$$I = A - P = 104.32 - 100 = \$4.32$$

So the compounded interest is \$4.32, which means the effective interest rate is 4.32%.

PRESENT VALUE

The **present value** of an investment is the original principal investment, the value of the investment before it earns any interest. Therefore, it is the principal P in our compound interest formula. But we can also use present value to determine how much money must be invested today for an investment to have a specific value at a future date.

Our compound interest formula states that $A = P\left(1 + \frac{r}{n}\right)^{nt}$. Solving for P, we get $P = \frac{A}{\left(1 + \frac{r}{n}\right)^{nt}}$.

The present value formula is $P = \dfrac{A}{\left(1+\dfrac{r}{n}\right)^{nt}}$ where P is the original principal (the present value), A is the compound amount, r is the APR, n is the number of compounding periods per year, and t is the number of years.

PROBLEM

The Hendricks wish to purchase a new car worth $20,000 for their newborn son, Charles, when he graduates high school at age 18. If they invest their money in an account compounded daily at 4.75% APR, how much should they invest when Charles is born?

SOLUTION

$$P = \frac{A}{\left(1+\frac{r}{n}\right)^{nt}} = \frac{20{,}000}{\left(1+\frac{0.0475}{360}\right)^{360(18)}} = \$8{,}506.14$$

They should invest $8,506.14. To do this calculation, it is easier to find the value of the denominator first and then divide that answer into 20,000.

PROBLEM

At age 25, Andrea received a large bonus from her employer. Rather than buy a house or a car, she decided to invest it. Her goal is to have $100,000 in a retirement account when she retires at age 65. If she gets an interest rate of 6.5% compounded monthly, how much should she invest?

SOLUTION

$$P = \frac{A}{\left(1+\frac{r}{n}\right)^{nt}} = \frac{100{,}000}{\left(1+\frac{0.065}{12}\right)^{12(40)}} = \$7{,}479.65$$

She should invest $7,479.65. Realize that because of inflation, in 40 years $100,000 won't have the same buying power as it does today.

INFLATION

When you invest money for a period of time in an interest-bearing account, you will have more money than you originally deposited. But that does not mean that you will be able to buy more with the compounded investment than you could have with the original investment. The reason is the effect of **inflation**. Inflation is an economic condition during which there are increases in the costs of goods and services. Inflation is usually expressed as an annual percentage rate. If the rate of inflation is greater than the compounding rate, an investor is actually losing money.

For example, suppose the price of a new computer is $2,000. You have enough money to purchase the computer, but decide to invest the $2,000 into an account paying 3.5% compounded quarterly. After one year, the compound amount is $2,070.92. If the rate of inflation is 4.5%, the cost of the computer is 2,000(1.045) = $2,090. Because $2,070.92 < $2,090, you have lost purchasing power. The compounded amount is not enough to pay for that same computer. Your money lost value because it buys less than it could one year before.

PROBLEM

You currently make $40,000. You want to know what an equivalent salary will be in 15 years if the inflation rate is 1.75%. You are finding a salary that will have the same purchasing power as today's salary.

SOLUTION

$P = 40{,}000$, $r = 0.0{,}175$, $n = 1$ (since inflation is an annual rate), $t = 15$

$$A = P\left(1+\frac{r}{n}\right)^{nt} = 40{,}000\left(1+\frac{0.0175}{1}\right)^{1(15)} = \$51{,}889.11$$

15 years from now, you need to earn an annual salary of $51,889.11 to have the same purchasing power as today.

PROBLEM

In the year 2020, you purchase an insurance policy that will pay you $500,000 when you retire in the year 2050. Assuming an annual inflation rate of 3%, what will be the purchasing power of the $500,000 in the year 2050?

SOLUTION

$A = 500{,}000$, $r = 0.03$, $n = 1$ (since inflation is an annual rate), $t = 30$

$$P = \frac{A}{\left(1+\frac{r}{n}\right)^{nt}} = \frac{500000}{\left(1+\frac{0.03}{1}\right)^{1(30)}} = 205{,}993.38$$

The purchasing power of half a million dollars in 2050 is only \$205,993.38.

Drill Questions

1. 12 is what percent of 25?

   (A) 40
   (B) 48
   (C) 3
   (D) 300

2. 40 is 8% of what number?

   (A) 32
   (B) 3.2
   (C) 50
   (D) 500

3. Which of the following has the largest value?

   (A) 90% of 90%
   (B) 120% of 80%
   (C) 20% of 500%
   (D) 0.5% of 2,000%

4. Frederick purchased an airplane for $79,500. He sold the plane three years later for $92,000. If the capital gains tax rate is 15%, what is the capital gains tax on the sale of the plane?

   (A) $1,875
   (B) $5,850
   (C) $11,925
   (D) $13,800

5. Bernie is a single taxpayer who made $1 million this year. He pays $119,996 + 39.6% of the amount over $413,200 in taxes. What percentage of his income goes to taxes?

   (A) 28.3%
   (B) 31.7%
   (C) 35.2%
   (D) 39.6%

6. In his first job, Steve paid $1,600 in federal taxes. He is in the bracket that requires him to pay $923 + 15% of the amount his yearly pay is over $9,225. What is his yearly pay?

(A) $10,667
(B) $11,531
(C) $12,745
(D) $13,740

7. Ray received $13,550 for the trade-in of his car. He purchased a new car for $45,200. If the state charges 6% sales tax and $75 for tags, what is his total payment for the new car?

(A) $30,912
(B) $33,624
(C) $34,362
(D) $34,437

8. A car dealership advertises that for one day only, the cost of a $15,000 car will be reduced to $14,000. What is the percent decrease in price?

(A) 2.1%
(B) 6.7%
(C) 7.1%
(D) 93.3%

9. The list price for a men's suit is $350. The store bought it for $200. What is the store's percentage markup?

(A) 64.6%
(B) 57.1%
(C) 42.9%
(D) 75%

10. An electronics store purchases a television for $1,000 and marks it up 30%. When the new models come out, they discount the price of the old models by 30%. What is the store's net on the sale of this TV?

(A) Store broke even
(B) Store lost $90
(C) Store made $30
(D) Store lost $300

11. Furniture store A discounts a sofa 50% while furniture store B discounts the same sofa 30% and then 20%. Which store offers the best deal?

    (A) Store A
    (B) Store B
    (C) They give the same deal.
    (D) It depends on the price of the sofa.

12. John's parents lend him $12,000 to buy a used car that he can drive to work. He promises to pay them back in 200 days at 3% simple interest. How much will he pay them back?

    (A) $200
    (B) $12,036
    (C) $12,200
    (D) $12,360

13. What is the approximate interest paid on a 5-year $15,000 certificate of deposit at 4.25% APR compounded quarterly?

    (A) $3,531
    (B) $3,470
    (C) $18,470
    (D) $18,531

14. To the nearest $50, what is the dollar difference in the future value of an investment of $25,000 over 10 years at 3.4% compounded annually and compounded daily?

    (A) $50
    (B) $100
    (C) $200
    (D) $9,250

15. You take out a loan for $50,000 and plan to pay it back in 8 years. You have a choice of different payback plans. Which one is the most advantageous to you?

    (A) 5.2% at simple interest
    (B) 4.5% compounded annually
    (C) 4.4% compounded semi-annually
    (D) 4.3% compounded monthly

16. Different banks offer nominal interest rates for their large investors. Which bank gives the highest effective interest rate?

    (A) 5.9% compounded annually
    (B) 5.8% compounded quarterly
    (C) 5.7% compounded monthly
    (D) 5.6% compounded continuously

17. Diane is 40 years old and wins $100,000 in a lottery after taxes. She decides to spend some on a car and a vacation and to invest the rest. She wants the amount that she invests to grow to the value of $100,000 at the time of her early retirement at age 62. How much money would she have to spend on the car and vacation if she invests her money at 4.8% compounded daily?

    (A) $34,787
    (B) $5,101
    (C) $94,899
    (D) $65,213

18. Ian puts the same amount of money into two different banks, each with a variable rate account. Bank A gives 4% APR the first year and then 3% APR the next year, each compounded quarterly. Bank B gives 3% APR the first year and 4% APR the second year, each compounded quarterly. Which bank gives a higher future value?

    (A) Bank A
    (B) Bank B
    (C) They both have the same future value.
    (D) It depends on the principal.

19. The cost of medical school is approximately $70,000 a year. A medical student has to borrow the money to pay for it. For each of 4 years, he borrows $70,000 at 4% interest compounded monthly. At the end of the four years, how much will he have to pay back?

    (A) $280,000
    (B) $291,200
    (C) $309,705
    (D) $328,496

20. A gallon of milk costs \$3.83. If the inflation rate stays at 4.3%, how much more will a gallon of milk cost 10 years from now?

(A) \$0.17
(B) \$2.01
(C) \$4.00
(D) \$5.84

Answers to Drill Questions

1. **(B)** $12 = \frac{x}{100}(25) \Rightarrow 25x = 1{,}200 \Rightarrow x = 48$

2. **(D)** $40 = \frac{8}{100}x \Rightarrow 8x = 4{,}000 \Rightarrow x = 500$

3. **(C)**

   (A) 09(0.9) = 0.81 = 81%

   (B) 1.2 (0.8) = 0.96 = 96%

   (C) 0.2(5) = 1 = 100%

   (D) 0.005 (20) = 0.1 = 10%

4. **(A)** 0.15(92,000 − 79,500) = 0.15(12,500) = \$1,875

5. **(C)**

   Tax = 119,996 + 0.396(1,000,000 − 413,200) = 119,996 + 0.396(586,800) = 119,996 + 232,373 = 352,369

   Percent in taxes: $\frac{352,369}{1,000,000} = 35.2\%$

6. **(D)**

   $\text{Tax} = 923 + 0.15(x = 9{,}225) = 1{,}600$

   $923 + 0.15x - 1383.75 = 1{,}600 = 0.15x = 2{,}060.75$

   $x = \frac{2{,}060.75}{0.15} = \$13{,}738.33$, which rounds to \$13,740.

   Note that this problem can also be done by trial and error.

7. **(B)**
Tax = 0.06(45,200 − 13,550) = 0.06(31,650) = \$1,899
Car cost = 31,650 + 1,899 + 75 = \$33,624

8. **(B)**

$$\%\text{ decrease} = \frac{\text{Original price} - \text{New price}}{\text{Original price}} = \frac{15{,}000 - 14{,}000}{15{,}000}$$

$$= \frac{1{,}000}{15{,}000} = 0.667 = 6.7\%$$

9. **(D)** Markup = Selling price − Cost = 350 − 200 = 150

$$\text{Percentage markup} = \frac{\text{Markup}}{\text{Cost}} = \frac{150}{200} = 75\%$$

10. **(B)** Price = 1,000 + 0.30(1,000) = 1,000 + 300 = \$1,300
Discount = 0.30(1,300) = 390
Discounted price = 1,300 − 390 = \$910
Store: Discounted price − Cost = 910 − 1,000 = −\$90

11. **(A)** Suppose the sofa's price is \$1,000. Store A will sell it for \$500. Store B discounts it \$300, making the price \$700 and then discounts it \$140 making the price \$560. Store A is better for the buyer.
If the price is x, Store A sells it for $0.5x$. Store B sells it for $0.8(0.7x) = 0.56x$.

12. **(C)**

$$P = 12{,}000,\ r = 3\% = 0.03,\ t = \frac{200\text{ days}}{360\text{ days}} = \frac{200}{360}$$

$$A = P + P \cdot r \cdot t = 12{,}000 + 12{,}000(0.03)\left(\frac{200}{360}\right) = 12{,}000 + 200 = \$12{,}200$$

13. **(A)**

$P = 15{,}000, r = 0.0{,}425, n = 4, t = 5$

$A = 15{,}000\left(1+\frac{0.0425}{4}\right)^{4(5)} = \$18{,}530.71$

$I = A - P = 18{,}530.71 - 15{,}000 = 3{,}530.71 \approx \$3{,}531$

14. **(B)**

Annually: $25{,}000\left(1+\frac{.034}{1}\right)^{1(10)} = \$34{,}925.72$

Daily: $25{,}000\left(1+\frac{.034}{360}\right)^{360(10)} = \$35{,}123.13$

Difference: $35{,}123.13 - 34{,}925.72 = 197.41 \approx \$197$

15. **(D)** With a loan, you want to pay back the least amount.

(A) $50{,}000 + 50{,}000(0.052)(8) = \$70{,}800$

(B) $50{,}000\left(1+\frac{.045}{1}\right)^{1(8)} = \$71{,}105.03$

(C) $50{,}000\left(1+\frac{.044}{2}\right)^{2(8)} = \$70{,}824.64$

(D) $50{,}000\left(1+\frac{.043}{12}\right)^{12(8)} = \$70{,}485.58$

16. **(B)**

(A) $100\left(1+\frac{.059}{1}\right)^{1} = 105.9$ (5.9% Effective interest)

(B) $100\left(1+\frac{.058}{4}\right)^{4} = \$105.93$ (5.93% Effective interest)

(C) $100\left(1+\frac{.057}{12}\right)^{12} = \$105.85$ (5.96% Effective interest)

(D) $100(2.71828)^{0.056} = \$105.76$ (5.76% Effective interest)

17. **(D)**

$A = 100{,}000,\ r = 0.048,\ n = 360,\ t = 22$

$$P = \frac{A}{\left(1+\frac{r}{n}\right)^{nt}} = \frac{100{,}000}{\left(1+\frac{0.048}{360}\right)^{360(22)}} = 34{,}786.89$$

She can spend about $\$100{,}000 - \$34{,}787 = \$65{,}213$ on the car and vacation.

18. **(C)**

Bank A: $P\left(1+\frac{.04}{4}\right)^{1(4)}\left(1+\frac{.03}{4}\right)^{1(4)}$

Bank B: $P\left(1+\frac{.03}{4}\right)^{1(4)}\left(1+\frac{.04}{4}\right)^{1(4)}$

These are the same no matter the value of P.

19. **(C)**

Year 1's money is borrowed for 4 years: $70{,}000\left(1+\frac{.04}{12}\right)^{12(4)} = \$82{,}124$

Year 2's money is borrowed for 3 years: $70{,}000\left(1+\frac{.04}{12}\right)^{12(3)} = \$78{,}909$

Year 3's money is borrowed for 2 years: $70{,}000\left(1+\frac{.04}{12}\right)^{12(2)} = \$75{,}820$

Year 4's money is borrowed for 1 year: $70{,}000\left(1+\frac{.04}{12}\right)^{12(1)} = \$72{,}852$

Total: $82{,}124 + 78{,}909 + 75{,}820 + 72{,}852 = \$309{,}705$

20. **(B)**

$P = 3.83,\ r = 0.043,\ n = 1$ (since inflation is an annual rate), $t = 10$

$$A = P\left(1+\frac{r}{n}\right)^{nt} = 3.83\left(1+\frac{0.043}{1}\right)^{1(10)} = \$5.84$$

$5.84 - 3.83 = \$2.01$

CHAPTER 7

Geometry Topics

CHAPTER 7

GEOMETRY TOPICS

Plane geometry refers to two-dimensional shapes (that is, shapes that can be drawn on a sheet of paper), such as triangles, parallelograms, trapezoids, and circles. Three-dimensional objects (that is, shapes with depth) are the subjects of solid geometry.

TRIANGLES

A closed three-sided geometric figure is called a **triangle**. The points of the intersection of the sides of a triangle are called the **vertices** of the triangle.

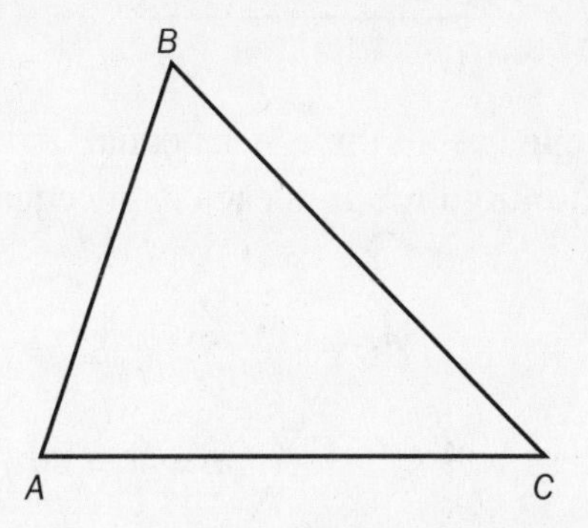

A **side** of a triangle is a line segment whose endpoints are the vertices of two angles of the triangle. The perimeter of a triangle is the sum of the measures of the sides of the triangle.

An **interior angle** of a triangle is an angle formed by two sides and includes the third side within its collection of points. The sum of the measures of the interior angles of a triangle is 180°.

A **scalene triangle** has no equal sides.

An **isosceles triangle** has at least two equal sides. The third side is called the **base** of the triangle, and the base angles (the angles opposite the equal sides) are equal.

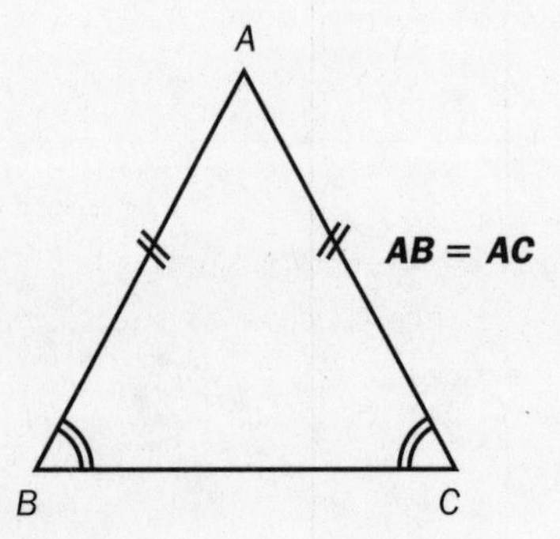

An **equilateral triangle** has all three sides equal. $\overline{AB} = \overline{AC} = \overline{BC}$. An equilateral triangle is also **equiangular**, with each angle equaling 60°.

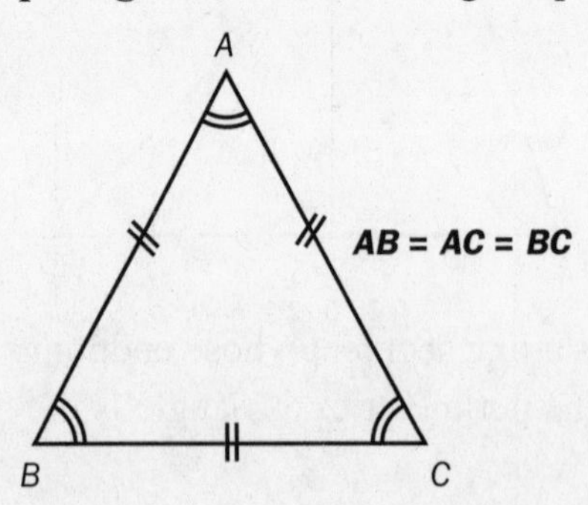

An **acute triangle** has three acute angles (less than 90°).

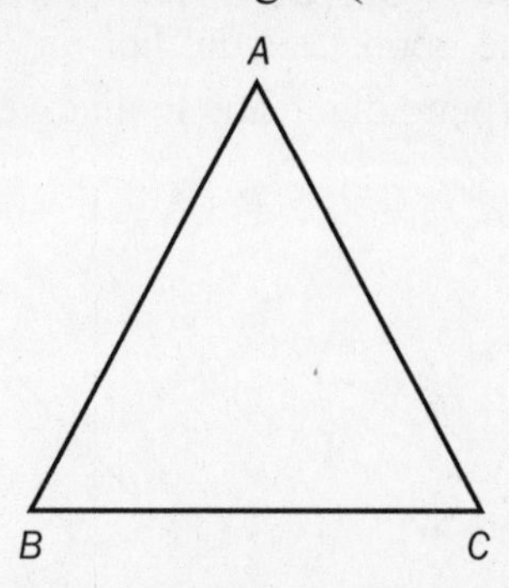

An **obtuse triangle** has one obtuse angle (greater than 90°).

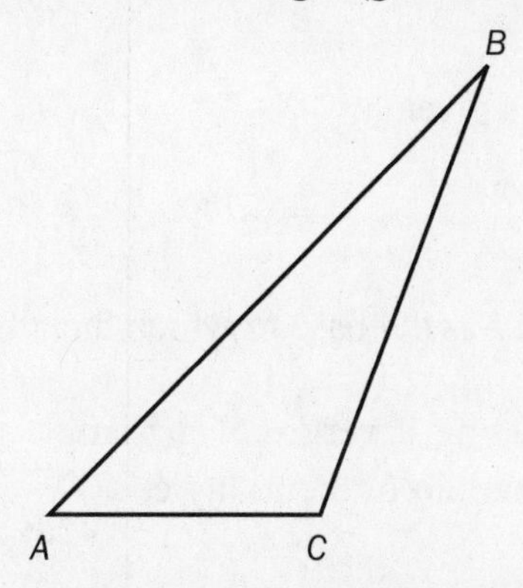

A **right triangle** has a right angle. The side opposite the right angle in a right triangle is called the **hypotenuse** of the right triangle. The other two sides are called the **legs** (or arms) of the right triangle. By the **Pythagorean Theorem**, the lengths of the three sides of a right triangle are related by the formula

$$c^2 = a^2 + b^2$$

where c is the hypotenuse and a and b are the other two sides (the legs). The Pythagorean Theorem is discussed in more detail in the next section.

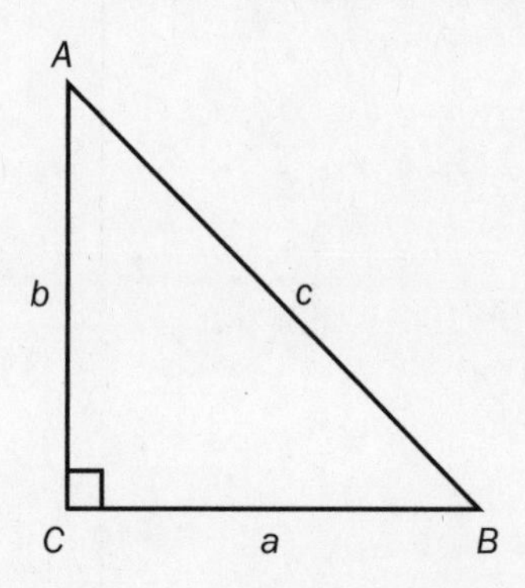

An **altitude**, or **height**, of a triangle is a line segment from a vertex of the triangle perpendicular to the opposite side. For an obtuse triangle, the altitude sometimes is drawn as a perpendicular line to an extension of the opposite side.

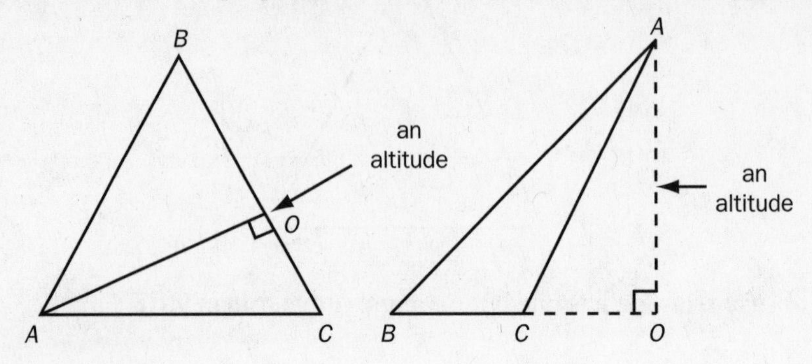

The **area** of a triangle is given by

$$A = \frac{1}{2}bh$$

where h is the altitude and b is the base to which the altitude is drawn.

A line segment connecting a vertex of a triangle and the midpoint of the opposite side is called a **median** of the triangle.

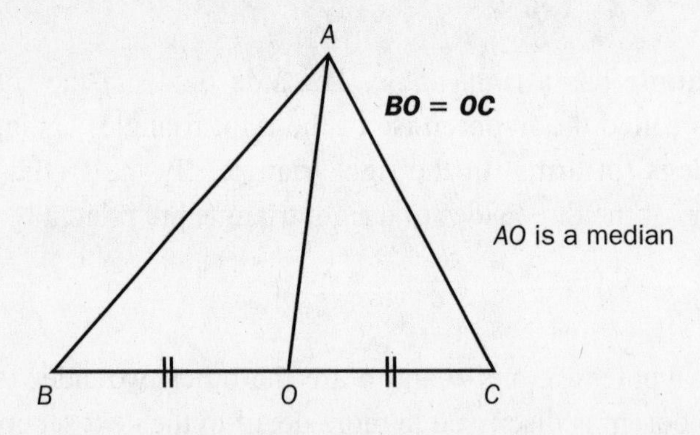

A line that bisects and is perpendicular to a side of a triangle is called a **perpendicular bisector** of that side.

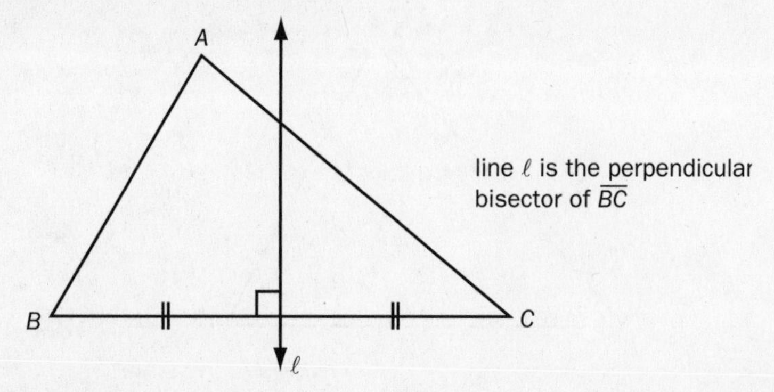

An **angle bisector** of a triangle is a line that bisects an angle and extends to the opposite side of the triangle.

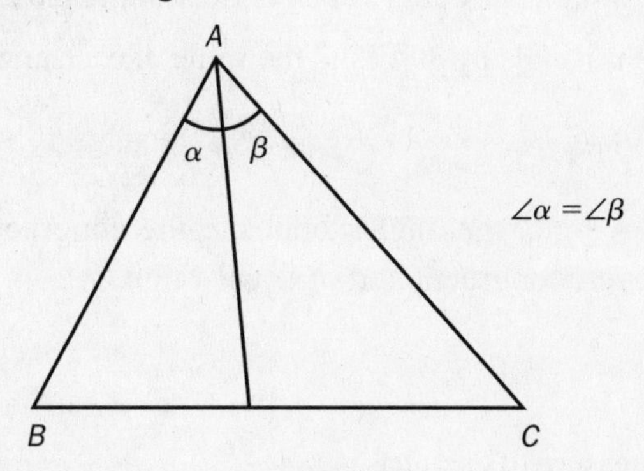

The line segment that joins the midpoints of two sides of a triangle is called a **midline** of the triangle.

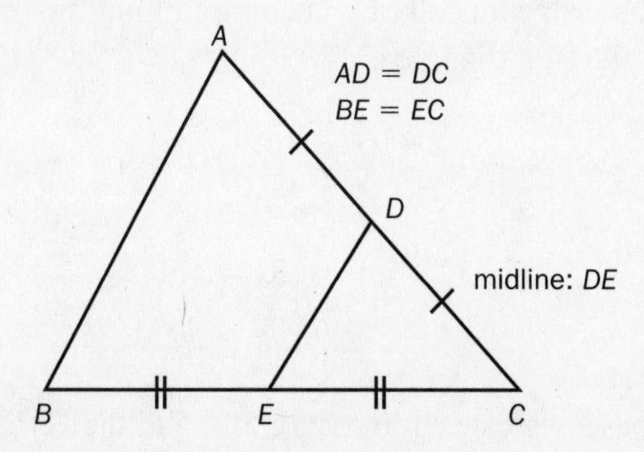

An **exterior angle** of a triangle is an angle formed outside a triangle by one side of the triangle and the extension of an adjacent side.

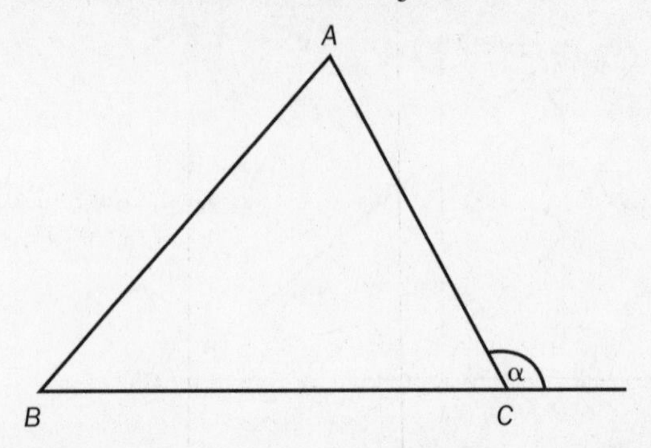

PROBLEM

The measure of the vertex angle of an isosceles triangle exceeds the measure of each base angle by 30°. Find the value of each angle of the triangle.

SOLUTION

In an isosceles triangle, the angles opposite the congruent sides (the base angles) are, themselves, congruent and of equal value.

Therefore,

1. Let x = the measure of each base angle
2. Then $x + 30$ = the measure of the vertex angle

We can solve for x algebraically by keeping in mind that the sum of all the measures of the angles of a triangle is 180°.

$$x + x + (x + 30) = 180$$
$$3x + 30 = 180$$
$$3x = 150$$
$$x = 50$$

Therefore, the base angles each measure 50°, and the vertex angle measures 80°.

THE PYTHAGOREAN THEOREM

The **Pythagorean Theorem** pertains to a right triangle, which, as we saw, is a triangle that has one 90° angle. The Pythagorean Theorem tells you that the square of the hypotenuse of a right triangle is equal to the sum of the squares of the other two sides, or

$$c^2 = a^2 + b^2$$

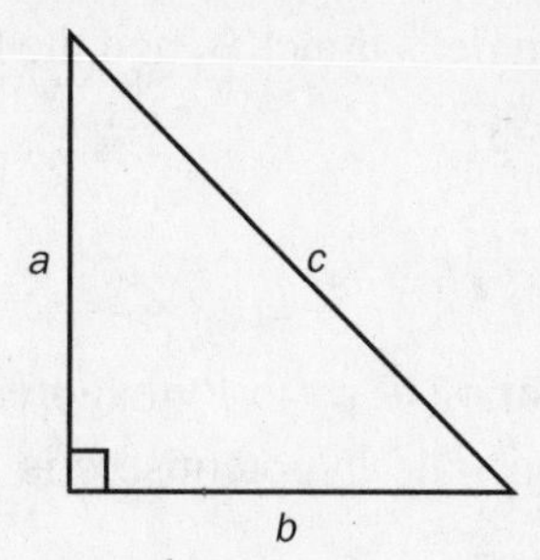

The Pythagorean Theorem is useful because if you know the length of any two sides of a right triangle, you can figure out the length of the third side.

PROBLEM

In a right triangle, one leg is 3 inches and the other leg is 4 inches. What is the length of the hypotenuse?

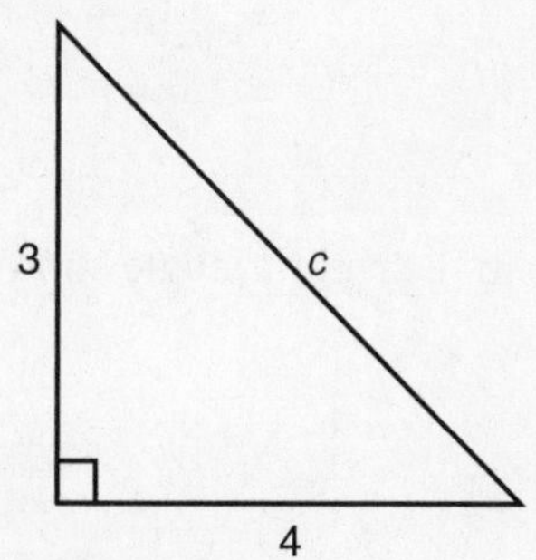

SOLUTION

$$c^2 = a^2 + b^2$$
$$c^2 = 3^2 + 4^2$$
$$c^2 = 9 + 16$$
$$c^2 = 25$$
$$c = 5$$

PROBLEM

If one leg of a right triangle is 6 inches, and the hypotenuse is 10, what is the length of the other leg?

SOLUTION

First, write down the equation for the Pythagorean Theorem. Next, plug in the information you are given. The hypotenuse c is equal to 10 and one of the legs, b, is equal to 6. Solve for a.

$$c^2 = a^2 + b^2$$
$$a^2 = c^2 - b^2$$
$$a^2 = 10^2 - 6^2$$
$$a^2 = 100 - 36$$
$$a^2 = 64$$
$$a = 8 \text{ inches}$$

PROBLEM

What is the value of b in the right triangle shown below?

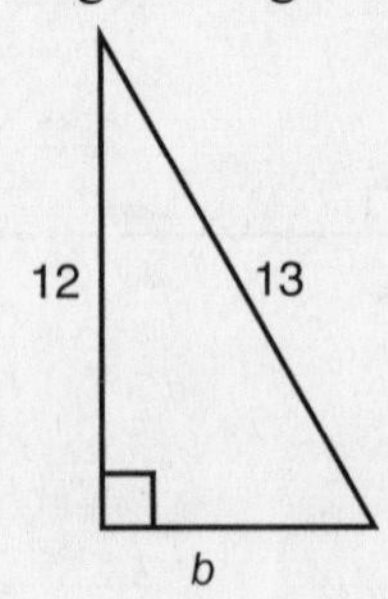

SOLUTION

To answer this question, you need to use the Pythagorean Theorem. The problem is asking for the value of the missing leg.

$$c^2 = a^2 + b^2$$
$$b^2 = c^2 - a^2$$
$$b^2 = 13^2 - 12^2$$
$$b^2 = 169 - 144$$
$$b^2 = 25$$
$$b = 5$$

QUADRILATERALS

A **polygon** is any closed figure with straight line segments as sides. A **quadrilateral** is any polygon with four sides. The points where the sides meet are called **vertices** (singular: **vertex**).

PARALLELOGRAMS

A **parallelogram** is a quadrilateral whose opposite sides are parallel.

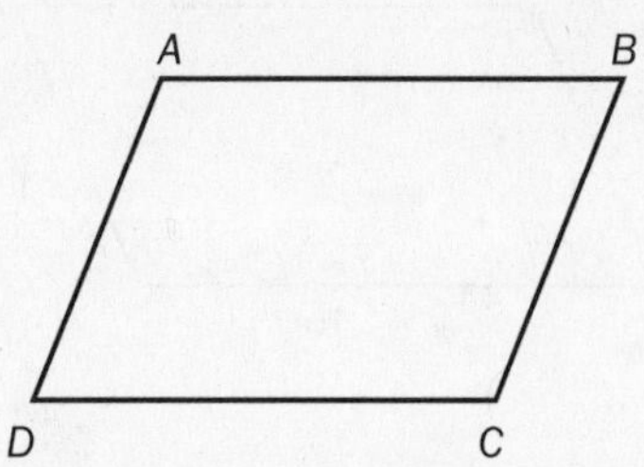

Two angles that have their vertices at the endpoints of the same side of a parallelogram are called **consecutive angles**. So $\angle A$ is consecutive to $\angle B$; $\angle B$ is consecutive to $\angle C$; $\angle C$ is consecutive to $\angle D$; and $\angle D$ is consecutive to $\angle A$.

The perpendicular segment connecting any point of a line containing one side of a parallelogram to the line containing the opposite side of the parallelogram is called the **altitude** of the parallelogram.

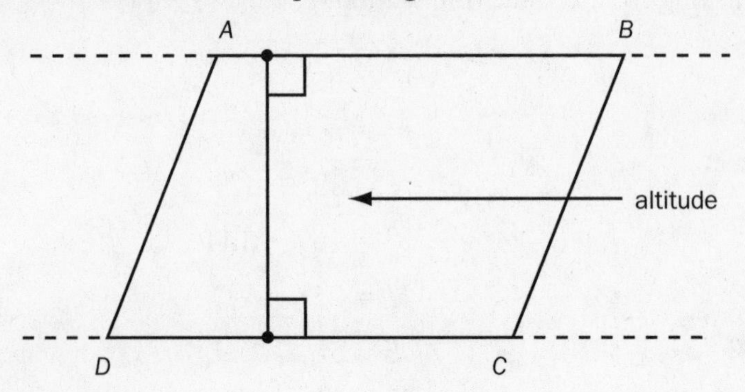

A **diagonal** of a polygon is a line segment joining any two nonconsecutive vertices. The area of a parallelogram is given by the formula $A = bh$, where b is the base and h is the height drawn perpendicular to that base. Note that the height is the same as the altitude of the parallelogram.

Example:

The area of the parallelogram below is:

$$A = bh$$

$$A = (10)(3)$$

$$A = 30$$

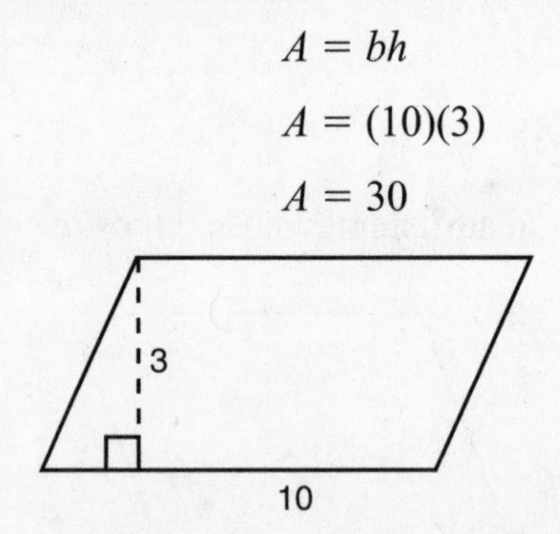

RECTANGLES

A **rectangle** is a parallelogram with right angles.

- The diagonals of a rectangle are equal, $\overline{AC} = \overline{BD}$.
- If the diagonals of a parallelogram are equal, the parallelogram is a rectangle.
- If a quadrilateral has four right angles, then it is a rectangle.
- The area of a rectangle is given by the formula $A = lw$, where l is the length and w is the width.

Example:

The area of the rectangle below is:

$$A = lw$$

$$A = (4)(9)$$

$$A = 36$$

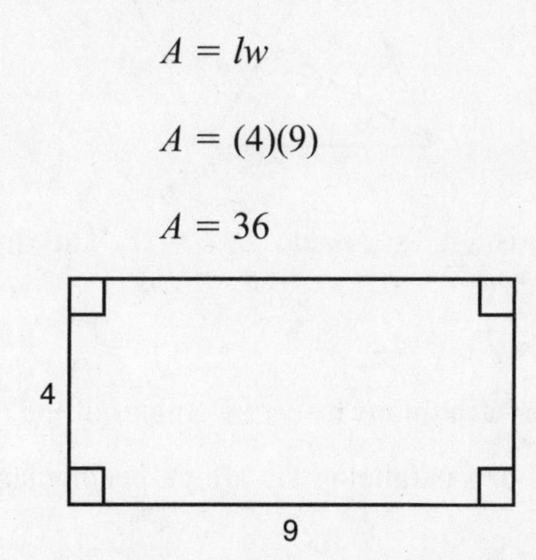

RHOMBI

A **rhombus** (plural: **rhombi**) is a parallelogram that has two adjacent sides that are equal.

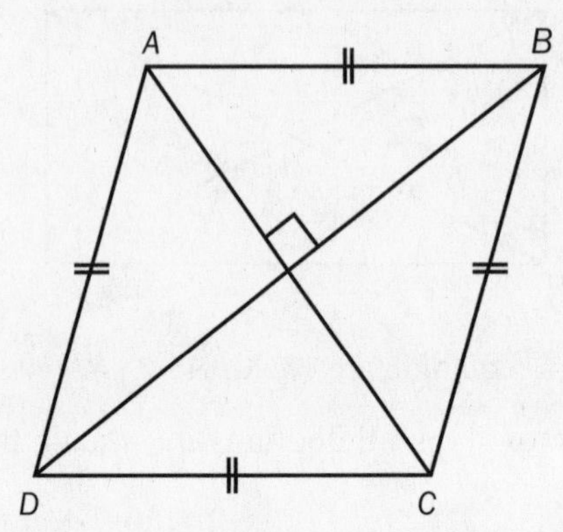

- All sides of a rhombus are equal.
- The diagonals of a rhombus are perpendicular bisectors of each other.
- The area of a rhombus can be found by the formula $A = \frac{1}{2}(d_1 \times d_2)$, where d_1 and d_2 are the diagonals.

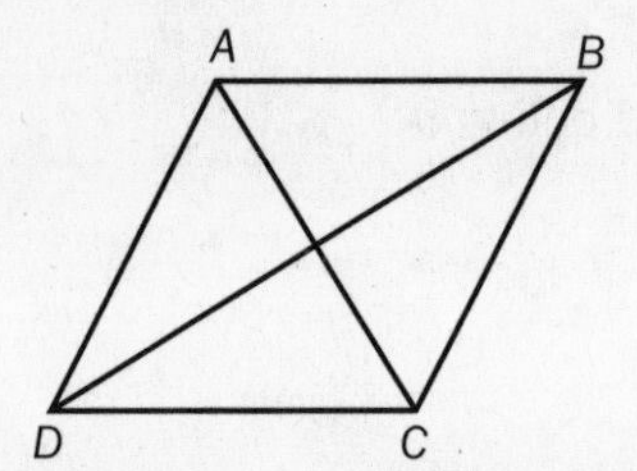

ABCD is a rhombus. $AC = 4$ and $BD = 7$. The area of the rhombus is $\left(\frac{1}{2}\right)(AC)(BD) = \left(\frac{1}{2}\right)(4)(7) = 14$.

- The diagonals of a rhombus bisect the angles of the rhombus.
- If the diagonals of a parallelogram are perpendicular, the parallelogram is a rhombus.
- If a quadrilateral has four equal sides, then it is a rhombus.
- A parallelogram is a rhombus if either diagonal of the parallelogram bisects the angles of the vertices it joins.

SQUARES

A **square** is a rhombus with a right angle.

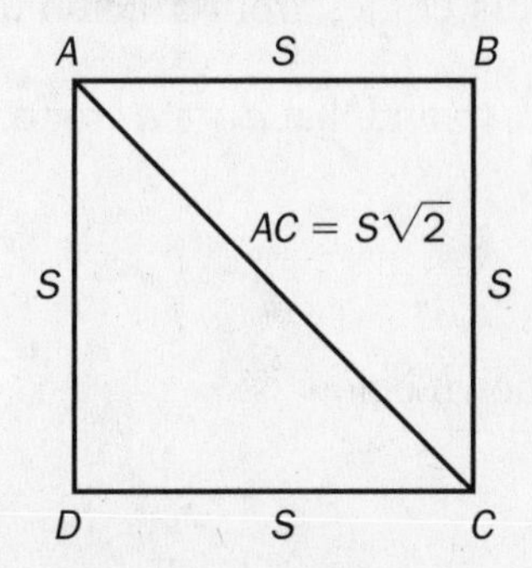

- A square is an equilateral quadrilateral.
- A square has all the properties of rhombi and rectangles.
- In a square, the measure of either diagonal can be calculated by multiplying the length of any side by the square root of 2.
- The area of a square is given by the formula $A = s^2$, where s is the side of the square.
- Since all sides of a square are equal, it does not matter which side is used.

Example:

The area of the square shown below is:

$$A = s^2$$

$$A = 6^2$$

$$A = 36$$

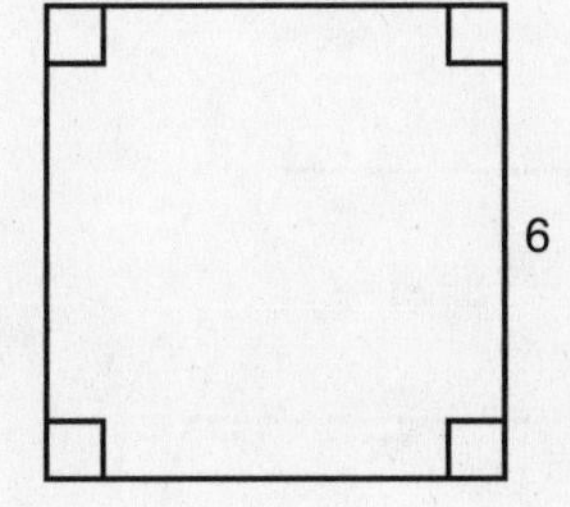

The area of a square can also be found by taking $\frac{1}{2}$ the product of the length of the diagonal squared. This comes from a combination of the facts that the area of a rhombus is $\left(\frac{1}{2}\right) d_1 d_2$ and that $d_1 = d_2$ for a square.

Example:

The area of the square shown below is:

$$A = \frac{1}{2}d^2$$

$$A = \frac{1}{2}(8)^2$$

$$A = 32$$

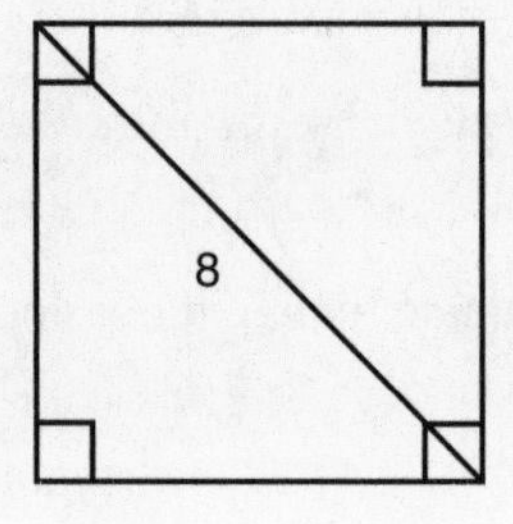

TRAPEZOIDS

A **trapezoid** is a quadrilateral with two and only two parallel sides. The parallel sides of a trapezoid are called the **bases**. The **median** of a trapezoid is the line joining the midpoints of the nonparallel sides.

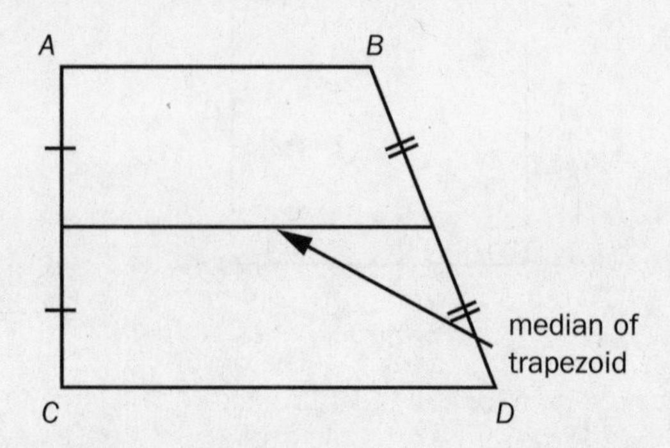

The perpendicular segment connecting any point in the line containing one base of the trapezoid to the line containing the other base is the **altitude** of the trapezoid.

A pair of angles including only one of the parallel sides is called a pair of **base angles**.

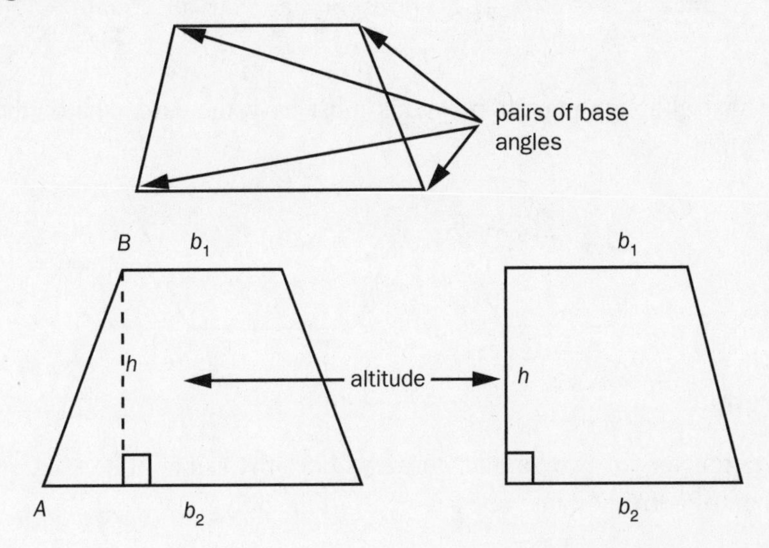

- The median of a trapezoid is parallel to the bases and equal to one-half their sum.
- The area of a trapezoid equals one-half the altitude times the sum of the bases, or $\frac{1}{2}h(b_1 + b_2)$.
- An **isosceles trapezoid** is a trapezoid whose non-parallel sides are equal. A pair of angles including only one of the parallel sides is called a pair of base angles.

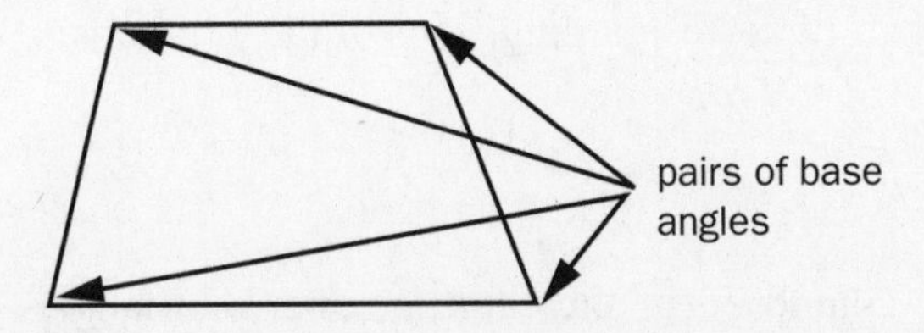

- The base angles of an isosceles trapezoid are equal.
- The diagonals of an isosceles trapezoid are equal.
- The opposite angles of an isosceles trapezoid are supplementary.

SIMILAR POLYGONS

Two polygons are **similar** if there is a one-to-one correspondence between their vertices such that all pairs of corresponding angles are congruent and the ratios of the measures of all pairs of corresponding sides are equal.

Note that although similar polygons must have the same shape, they may have different sizes.

Theorem 1

The perimeters of two similar polygons have the same ratio as the measure of any pair of corresponding line segments of the polygons.

Theorem 2

The ratio of the lengths of two corresponding diagonals of two similar polygons is equal to the ratio of the lengths of any two corresponding sides of the polygons.

Theorem 3

The areas of two similar polygons have the same ratio as the square of the measures of any pair of corresponding sides of the polygons.

Theorem 4

Two polygons composed of the same number of triangles similar to each, and similarly placed, are similar. Thus, $ABCD$ is similar to $A'B'C'D'$. Note that when naming similar polygons, the corresponding letters must match: A to A', B to B', C to C', and D to D'.

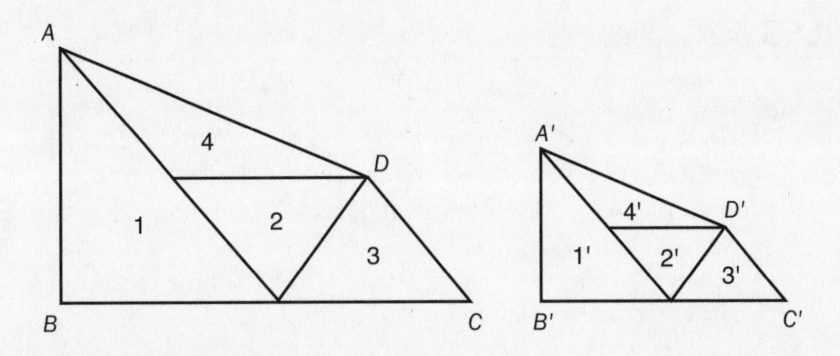

PROBLEM

The lengths of two corresponding sides of two similar polygons are 4 and 7. If the perimeter of the smaller polygon is 20, find the perimeter of the larger polygon.

SOLUTION

We know, by theorem, that the perimeters of two similar polygons have the same ratio as the measures of any pair of corresponding sides.

If we let s and p represent the side and perimeter of the smaller polygon and s' and p' represent the corresponding side and perimeter of the larger one, we can then write the proportion

$$s : s' = p : p'; \text{ or } \frac{s}{s'} = \frac{p}{p'}$$

By substituting the given values, we can solve for p'.

$$\frac{4}{7} = \frac{20}{p'}$$

$$4p' = 140$$

$$p' = 35$$

Therefore, the perimeter of the larger polygon is 35.

CIRCLES

A **circle** is a set of points in the same plane equidistant from a fixed point, called its **center**. Circles are often named by their center point, such as circle ***O*** below.

A **radius** of a circle is a line segment drawn from the center of the circle to any point on the circle.

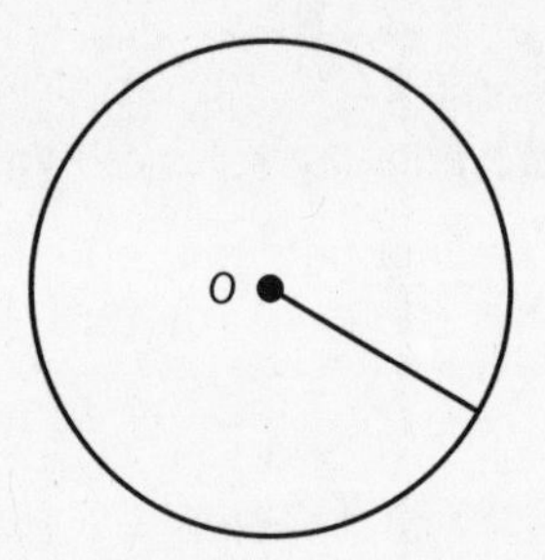

The **circumference** of a circle is the length of its outer edge, given by

$$C = \pi d = 2\pi r$$

where r is the radius, d is the diameter, and π (pi) is a mathematical constant approximately equal to 3.14.

The **area** of a circle is given by

$$A = \pi r^2$$

A full circle is 360°. The measure of a semicircle (half a circle) is 180°.

A line that intersects a circle in two points is called a **secant**.

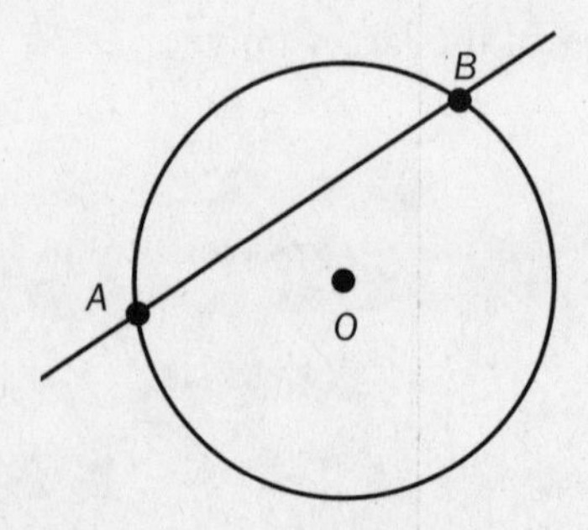

A line segment joining two points on a circle is called a **chord** of the circle.

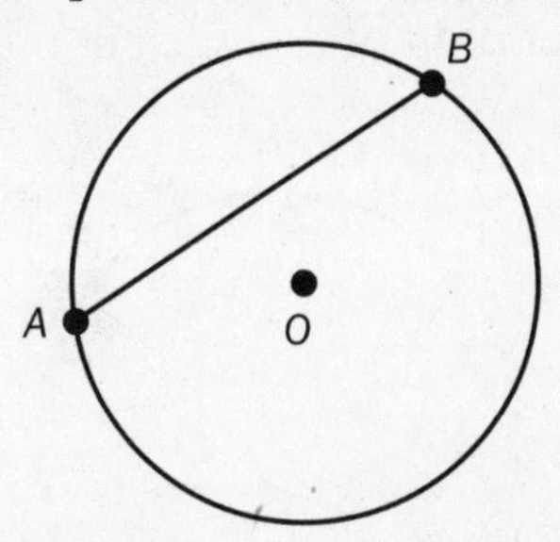

A chord that passes through the center of the circle is called a **diameter** of the circle. The length of the diameter is twice the length of the radius, $d = 2r$.

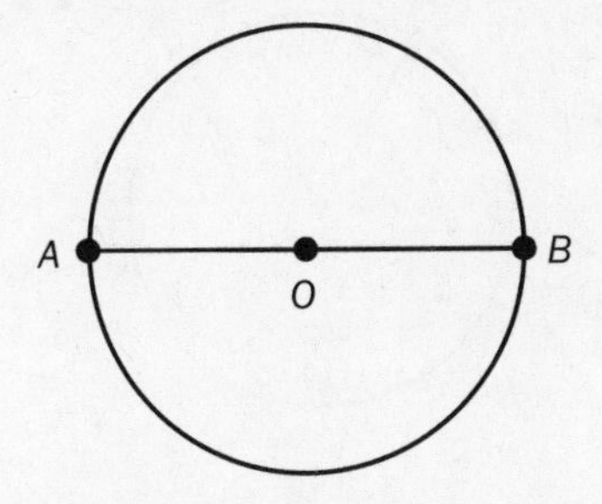

The line passing through the centers of two (or more) circles is called the **line of centers**.

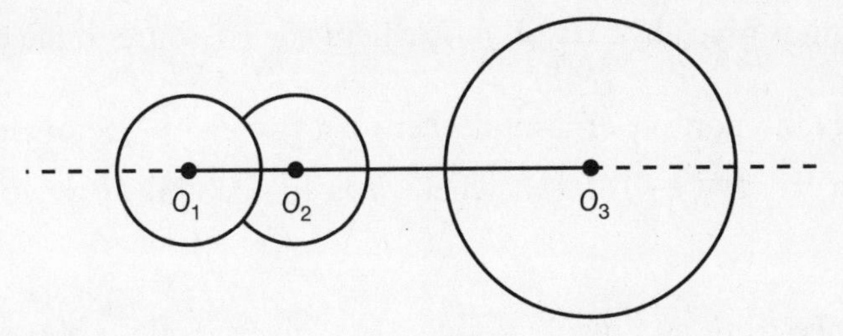

An angle whose vertex is on the circle and whose sides are chords of the circle is called an **inscribed angle** ($\angle BAC$ in the diagrams).

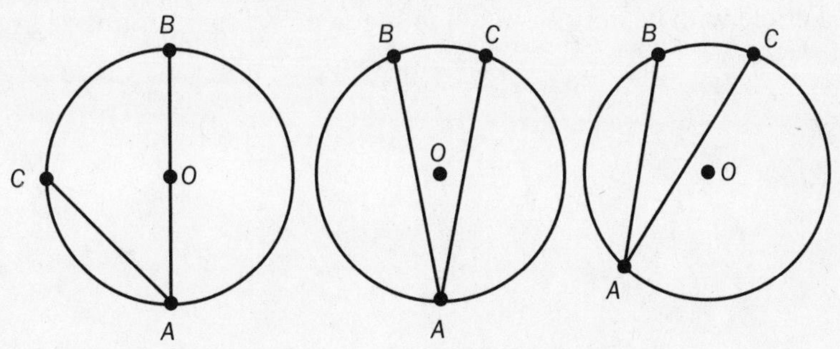

An angle whose vertex is at the center of a circle and whose sides are radii is called a **central angle**. The portion of a circle cut off by a central angle is called an **arc** of the circle.

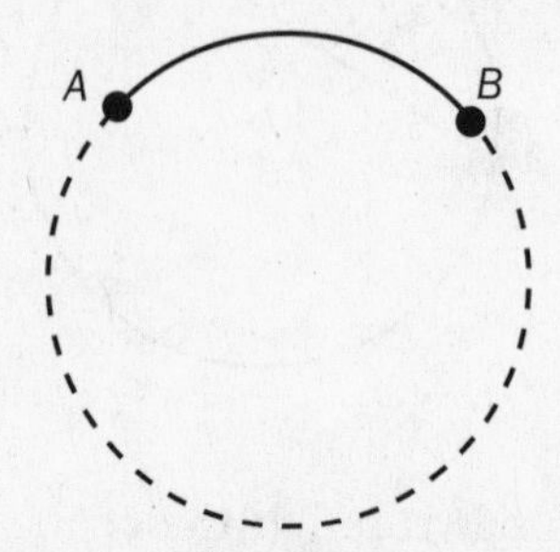

The measure of a minor arc is the measure of the central angle that intercepts that arc. The measure of a semicircle (half a circle) is 180°.

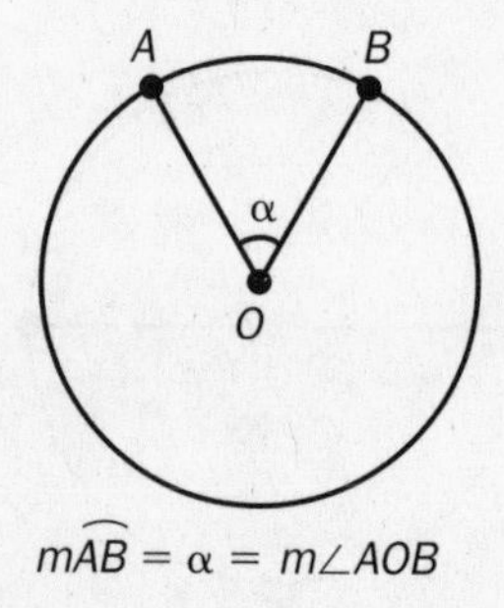

$m\widehat{AB} = \alpha = m\angle AOB$

The length of an arc intercepted by a central angle has the same ratio to the circle's circumference as the measure of the arc is to 360°, the full circle. Therefore, arc length is given by $\frac{n}{360} \times 2\pi r$, where n = measure of the central angle.

A sector is the portion of a circle between two radii (sector AOB here). Its area is given by $A = \frac{n}{360}(\pi r^2)$, where n is the central angle formed by the radii.

The distance from an outside point P to a given circle is the distance from that point to the point where the circle intersects with a line segment with endpoints at the center of the circle and point P. The distance of point P to the diagrammed circle with center O is the line segment $\overline{PB}$, part of line segment $\overline{PO}$.

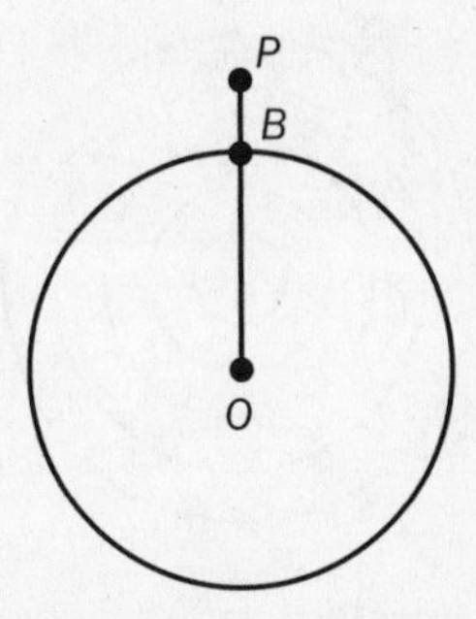

A line that has one and only one point of intersection with a circle is called a **tangent** to that circle, and their common point is called a **point of tangency**.

In the diagram, Q and P are each points of tangency. A tangent is always perpendicular to the radius drawn to the point of tangency.

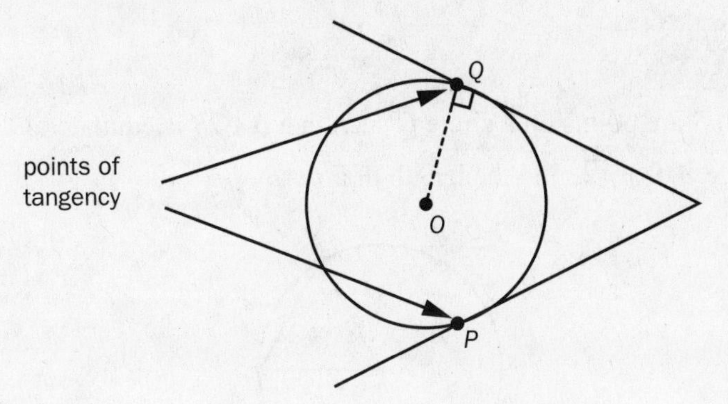

Congruent circles are circles whose radii are congruent.

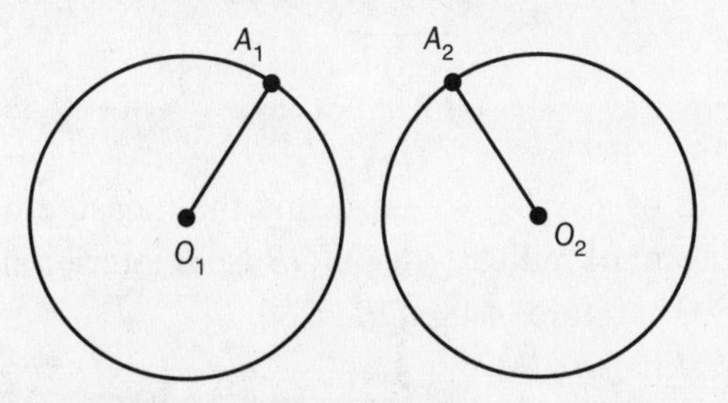

If $O_1A_1 \cong O_2A_2$, then $O_1 \cong O_2$.

Circles that have the same center and unequal radii are called **concentric circles**.

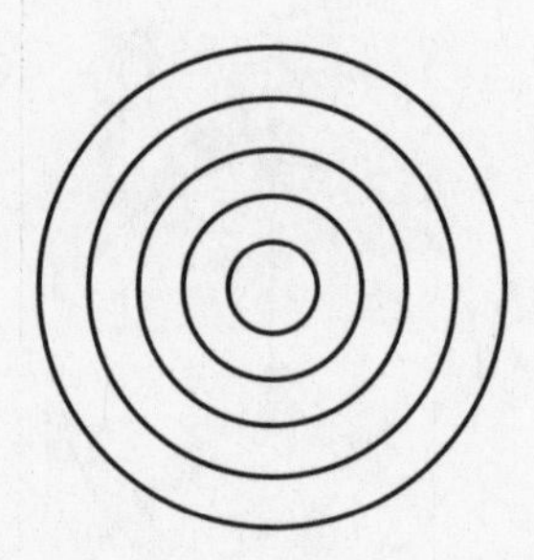

A **circumscribed circle** is a circle passing through all the vertices of a polygon. The polygon is said to be **inscribed** in the circle.

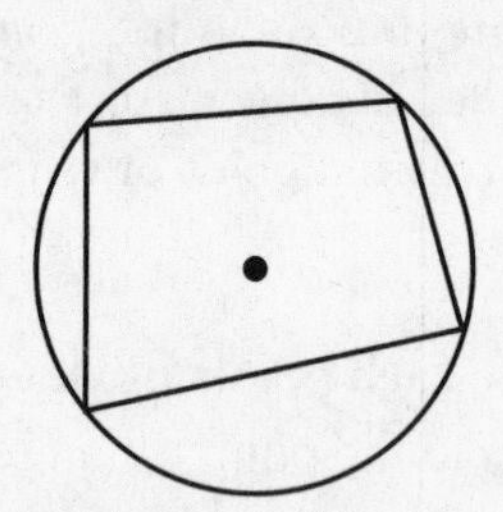

PROBLEM

A and B are points on a circle Q such that ΔAQB is equilateral. If the length of side $\overline{AB} = 12$, find the length of arc AB.

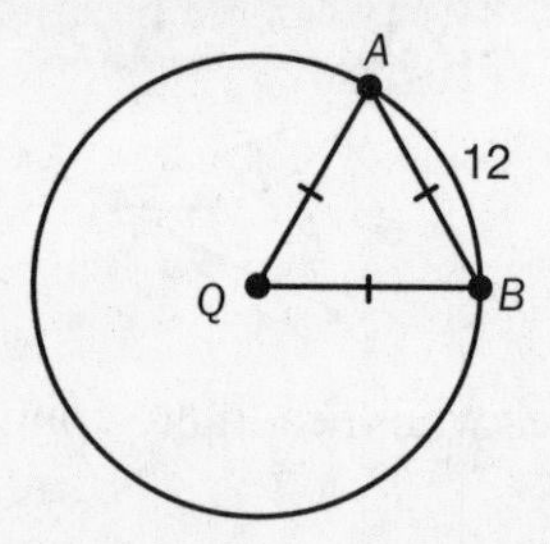

SOLUTION

To find the length of arc AB, we must find the measure of the central angle $\angle AQB$ and the measure of radius $\overline{QA}$. $\angle AQB$ is an interior angle of the equilateral triangle ΔAQB. Therefore, $m\ \angle AQB = 60°$.

Similarly, in the equilateral ΔAQB, $\overline{AQ} = \overline{AB} = \overline{QB} = 12 = r$.

Given the radius, r, and the central angle, n, the arc length is given by

$$\frac{n}{360} \times 2\pi r = \frac{60}{360} \times 2\pi \times 12 = \frac{1}{6} \times 2\pi \times 12 = 4\pi.$$

Therefore, the length of arc $AB = 4\pi$.

FORMULAS FOR AREA AND PERIMETER

Figures	**Areas**
Area (A) of a:	
square	$A = s^2$; where s = side
rectangle	$A = lw$; where l = length, w = width
parallelogram	$A = bh$; where b = base, h = height
triangle	$A = \frac{1}{2}bh$; where b = base, h = height
circle	$A = \pi r^2$; where $\pi = 3.14$, r = radius
sector	$A = \left(\frac{n}{360}\right)(\pi r^2)$; where n = central angle, r = radius, $\pi = 3.14$
trapezoid	$A = \left(\frac{1}{2}\right)(h)(b_1 + b_2)$; where h = height, b_1 and b_2 = bases

Figures	**Perimeters**
Perimeter (P) of a:	
square	$P = 4s$; where s = side
rectangle	$P = 2l + 2w$; where l = length, w = width
triangle	$P = a + b + c$; where a, b, and c are the sides
Circumference (C) of a circle	$C = \pi d$; where $\pi = 3.14$, d = diameter

PROBLEM

Points P and R lie on circle Q, $m\angle PQR = 120°$, and $PQ = 18$. What is the area of sector PQR?

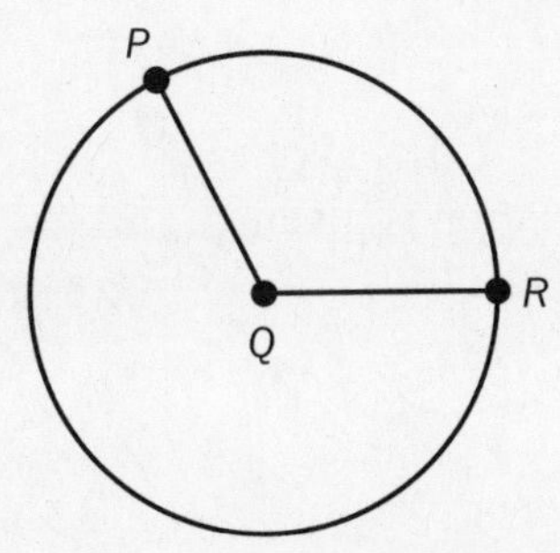

SOLUTION

$$\frac{120°}{360°} = \frac{\text{Area of sector } PQR}{\text{Area of circle } Q}$$

Letting X = area of sector PQR, and replacing area of circle Q with $\pi(18^2) = 324\pi$, we get

$$\frac{120°}{360°} = \frac{X}{324\pi}$$

$$\text{Then } X = \frac{(120°)(324\pi)}{360°} = 108\pi$$

Drill Questions

1. What is the area of the following right triangle?

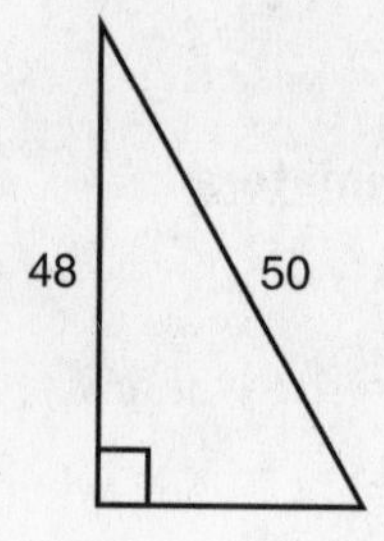

(A) 1200
(B) 672
(C) 336
(D) 112

2. Parallelogram $RSTU$ is similar to parallelogram $WXYZ$. If $\angle RST = 60°$, $\angle XYZ =$

(A) 60°
(B) 90°
(C) 120°
(D) Not enough information is given

3. The center of circle O is at the origin, as shown. Point (2, 2) is on the circle. What is the circumference of circle O?

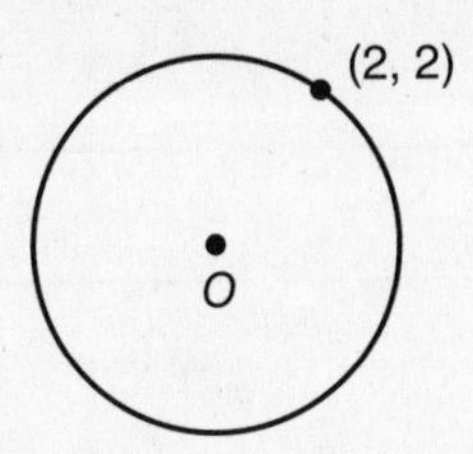

(A) 4π
(B) $4\sqrt{2}\pi$
(C) 2π
(D) $2\sqrt{2}\pi$

4. If the short side of a rectangle measures 3 inches, and its long side is twice as long, what is the length of a diagonal of a square with the same area as this rectangle?

(A) $3\sqrt{2}$
(B) 6
(C) 18
(D) 36

5. Which of the following *cannot* be the lengths of the sides of a triangle?

(A) 2, 5, 6
(B) 3, 4, 5
(C) 4, 5, 6
(D) 4, 5, 10

6. Find the length of the missing side in this right triangle.

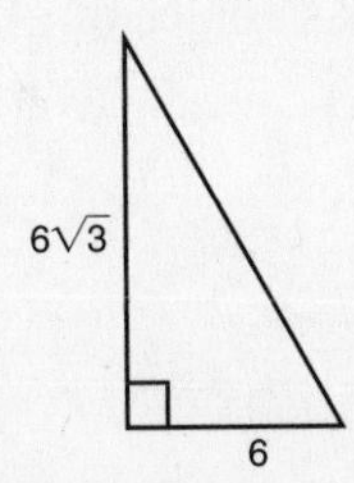

(A) 6
(B) $\sqrt{66}$
(C) 12
(D) $\sqrt{306}$

7. Given the intersecting lines and angle measurement in the figure, $x =$

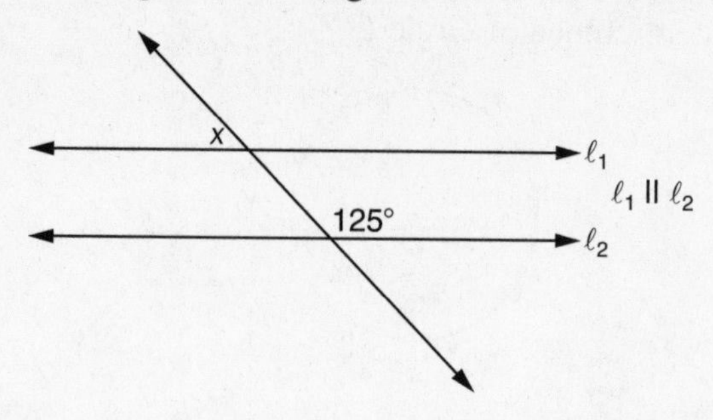

(A) 55°
(B) 60°
(C) 65°
(D) 70°

8. This figure shows sector *AOB* equal to a quarter of a circle, circumscribed about a square. What is the area of the shaded region?

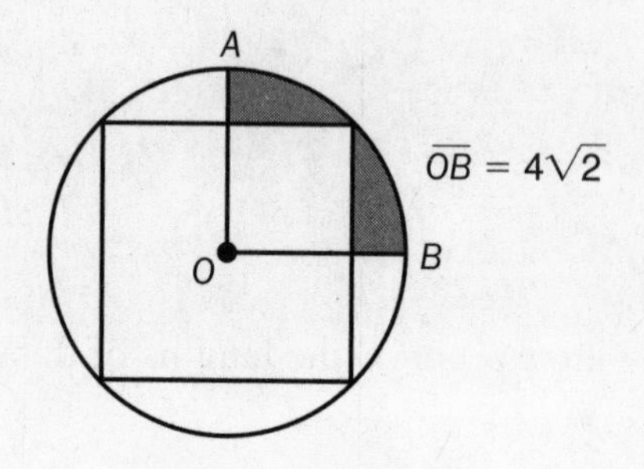

(A) $16(2\pi - 1)$
(B) $8\sqrt{2\pi} - 16$
(C) 8π
(D) $8(\pi - 2)$

9. An isosceles right triangle has angles of

(A) 30°, 60°, 90°
(B) 45°, 45°, 90°
(C) 0°, 90°, 90°
(D) 60°, 60°, 60°

10. An angle of measure of 180° is termed

    (A) straight
    (B) supplementary
    (C) obtuse
    (D) reflex

Answers to Drill Questions

1. **(C)** Let x represent the length of the horizontal base of this triangle. By the Pythagorean theorem, $x^2 + 48^2 = 50^2$. Then $x^2 + 2{,}304 = 2{,}500$, which means that $x^2 = 2{,}500 - 2{,}304 = 196$. So $x = \sqrt{196} = 14$. The area of the triangle is one-half the product of the base and the height. Thus, the required area is $\left(\frac{1}{2}\right)(14)(48) = \left(\frac{1}{2}\right)(672) = 336$.

2. **(C)** Corresponding angles of similar geometric figures must be congruent. This means that $\angle XYZ$ in parallelogram $WXYZ$ must be congruent to $\angle STU$ in parallelogram $RSTU$. We note that $\angle RST$ and $\angle STU$ are consecutive angles in $RSTU$ and the sum of any two consecutive angles in a parallogram is 180°. Since $\angle RST = 60°$, $\angle STU = 180° - 60° = 120°$. Thus, $\angle XYZ = 120°$.

3. **(B)** The length of the radius of this circle is found by the distance between point O which is located at (0, 0) and the point (2, 2). The distance between (0, 0) and (2, 2) is $\sqrt{(0-2)^2 + (0-2)^2} = \sqrt{(-2)^2 + (-2)^2} = \sqrt{4+4} = \sqrt{8}$. Note that we can write $\sqrt{8}$ as $\left(\sqrt{4}\right)\left(\sqrt{2}\right) = 2\sqrt{2}$. Finally, the circumference equals the product of 2π and the radius, which is $(2\pi)\left(2\sqrt{2}\right) = 4\pi\sqrt{2} = 4\sqrt{2}\pi$.

4. **(B)** The long side of the rectangle measures (2)(3) = 6 inches, so its area is (6)(3) = 18 square inches. Then the area of the square is also 18 square inches. Then each side of the square is $\sqrt{18}$ inches. The diagonal of a square can represent the hypotenuse of a triangle whose two legs are consecutive sides of the square. Let d represent the length of the diagonal. By the Pythagorean theorem, $d^2 = \left(\sqrt{18}\right)^2 + \left(\sqrt{18}\right)^2 = 18 + 18 = 36$. Thus, $d = \sqrt{36} = 6$.

5. **(D)** The sum of any two sides of a triangle must exceed the length of the third side. This means that a triangle *cannot* have sides with lengths of 4, 5, and 10 because $4 + 5 < 10$.

6. **(C)** The missing side represents the hypotenuse of this triangle. Let x represent its length. By the Pythagorean theorem, $x^2 + 6^2 + \left(6\sqrt{3}\right)^2$. This equation simplifies to $x^2 = 36 + (36)(3) = 144$. Thus, $x = \left(\sqrt{144}\right) = 12$.

7. **(A)** When a transversal intersects two parallel lines, corresponding angles are congruent. By definition, x and the angle represented by 125° are corresponding angles. As such, $x = 125°$. Now note that the angle represented by x and 125° are adjacent angles that form a straight line. This means that $x + 125° = 180°$. Thus, $x = 180° - 125° = 55°$.

8. **(D)** The diagonal of the square $\overline{OD}$ is a radius of the circle, so its length is equal to $\overline{OB}$, which is $4\sqrt{2}$. Let x represent the length of each side of the square. By the Pythagorean theorem, $\left(\overline{OC}\right)^2 + \left(\overline{CD}\right)^2 = \left(\overline{OD}\right)^2$. Then $x^2 + x^2 = \left(4\sqrt{2}\right)^2 = (16)(2) = 32$. This equation can be simplified to $2x^2 = 32$, which further simplifies to $x^2 = 16$. This means that the area of the square is 16. The area of the quarter-circle formed by the points A, O, and B is one-fourth the area of the circle. This area is found by the expression $\left(\frac{1}{4}\right)(\pi)\left(r^2\right)$, where r is the length of the radius. Since we know the radius length to be $4\sqrt{2}$, the area of the quarter-circle is $\left(\frac{1}{4}\right)(\pi)\left(4\sqrt{2}\right)^2 = \left(\frac{1}{4}\right)(\pi)(16)(2) = \left(\frac{1}{4}\right)(\pi)(32) = 8\pi$. Finally, the area of the shaded region is the difference between the area of the quarter-circle and the area of the square, which is $8\pi - 16$. This is equivalent to $8(\pi - 2)$.

9. **(B)** An isosceles right triangle must consist of a 90° angle and two acute congruent angles whose sum is 90°. The only set of numbers that satisfies these conditions are angle measures of 45°, 45°, 90°.

10. **(A)** By definition, an angle whose measure is 180° is called a straight angle.

CHAPTER 8

Logic and Sets

CHAPTER 8

LOGIC AND SETS

LOGIC

The topic of logic encompasses a wealth of subjects related to the principles of reasoning. Here, we will be concerned with logic on an elementary level, and you will recognize the principles introduced here because not only have you been using them in your everyday life, you also have seen them in one form or another in this book. The chapter is designed to familiarize you with the terminology you might be expected to know as well as thought processes that you can use in everyday life.

Sentential calculus is the "calculus of sentences," a field in which the truth or falseness of assertions is examined by using algebraic tools. We will approach logic from a "true" or "false" perspective here. This chapter contains many examples of sentences to illustrate the terms that are defined.

SENTENCES

> A **sentence** is any expression that can be labeled either true or false.

Examples:

Expressions to which the terms "true" or "false" can be assigned include the following:

1. "It is raining where I am standing."
2. "My name is George."
3. "1 + 2 = 3"

Examples:

Expressions to which the terms "true" or "false" cannot be assigned include the following:

1. "I will probably be healthier if I exercise."
2. "It will rain on this day, one year from now."
3. "What I am saying at this instant is a lie."

Sentences can be combined to form new sentences using the connectives **AND**, **OR**, **NOT**, and **IF-THEN**.

Examples:

The sentences

1. "John is tired."
2. "Mary is cooking."

. . .***can be combined to form***. . .

1. "John is tired AND Mary is cooking."
2. "John is tired OR Mary is cooking."
3. "John is NOT tired."
4. "IF John is tired, THEN Mary is cooking."

LOGICAL PROPERTIES OF SENTENCES

Consistency

A sentence is **consistent** if and only if it is *possible* that it is true. A sentence is **inconsistent** if and only if it is not consistent; that is, if and only if it is *impossible* that it is true.

Example:

"At least one odd number is not odd" is an inconsistent sentence.

Logical Truth

A sentence is **logically true** if and only if it is *impossible* for it to be false; that is, the denial of the sentence is inconsistent.

Example:

Either Mars is a planet or Mars is not a planet.

Logical Falsity

A sentence is **logically false** if and only if it is *impossible* for it to be true; that is, the sentence is inconsistent.

Example:

Mars is a planet and Mars is not a planet.

Logical Indeterminacy (Contingency)

A sentence is **logically indeterminate** (contingent) if and only if it is neither logically true nor logically false.

Example:

Einstein was a physicist and Pauling was a chemist.

Logical Equivalent of Sentences

Two sentences are **logically equivalent** if and only if it is *impossible* for one of the sentences to be true while the other sentence is false; that is, if and only if it is impossible for the two sentences to have different truth values.

Example:

"Chicago is in Illinois and Pittsburgh is in Pennsylvania" is logically equivalent to "Pittsburgh is in Pennsylvania and Chicago is in Illinois."

STATEMENTS

A **statement** is a sentence that is either true or false, but not both.

The following terms and their definitions should become familiar to you. Their logic is probably familiar, even though you haven't as yet given it a label.

CONJUNCTION

If a and b are statements, then a statement of the form "a and b" is called the **conjunction** of a and b, denoted by $a \wedge b$.

DISJUNCTION

The **disjunction** of two statements a and b is shown by the compound statement "a or b," denoted by $a \vee b$.

NEGATION

The **negation** of a statement q is the statement "not q," denoted by $\sim q$.

IMPLICATION

The compound statement "if a, then b," denoted by $a \rightarrow b$, is called a **conditional statement** or an **implication**. "If a" is called the **hypothesis** or **premise** of the implication, and "then b" is called the **conclusion** of the implication. Further, statement a is called the **antecedent** of the implication, and statement b is called the **consequent** of the implication.

CONVERSE

The **converse** of $a \to b$ is $b \to a$.

CONTRAPOSITIVE

The **contrapositive** of $a \to b$ is $\sim b \to \sim a$.

INVERSE

The **inverse** of $a \to b$ is $\sim a \to \sim b$.

BICONDITIONAL

The statement of the form "p if and only if q," denoted by $p \leftrightarrow q$, is called a **biconditional** statement.

VALIDITY

An argument is **valid** if the truth of the premises means that the conclusions must also be true.

INTUITION

Intuition is the process of making generalizations on insight.

PROBLEMS

Write the inverse for each of the following statements. Determine whether the inverse is true or false.

1. If a person is stealing, he is breaking the law.
2. If a line is perpendicular to a segment at its midpoint, it is the perpendicular bisector of the segment.
3. Dead men tell no tales.

SOLUTIONS

The inverse of a given conditional statement is formed by negating both the hypothesis and conclusion of the conditional statement.

1. The hypothesis of this statement is "a person is stealing"; the conclusion is "he is breaking the law." The negation of the hypothesis is "a person is not stealing." The inverse is "if a person is not stealing, he is not breaking the law."

   The inverse is false, since there are more ways to break the law than by stealing. Clearly, a murderer may not be stealing but he is surely breaking the law.

2. In this statement, the hypothesis contains two conditions: a) the line is perpendicular to the segment; and b) the line intersects the segment at the midpoint. The negation of (statement a *and* statement b) is (not statement a *or* not statement b). Thus, the negation of the hypothesis is "The line is not perpendicular to the segment or it doesn't intersect the segment at the midpoint." The negation of the conclusion is "the line is not the perpendicular bisector of a segment."

   The inverse is "if a line is not perpendicular to the segment or does not intersect the segment at the midpoint, then the line is not the perpendicular bisector of the segment."

   In this case, the inverse is true. If either of the conditions holds (the line is not perpendicular; the line does not intersect at the midpoint), then the line cannot be a perpendicular bisector.

3. This statement is not written in if-then form, which makes its hypothesis and conclusion more difficult to see. The hypothesis is implied to be "the man is dead"; the conclusion is implied to be "the man tells no tales." The inverse is, therefore, "If a man is not dead, then he will tell tales."

   The inverse is false. Many witnesses to crimes are still alive but they have never told their stories to the police, probably out of fear or because they didn't want to get involved.

BASIC PRINCIPLES, LAWS, AND THEOREMS

1. Any statement is either true or false. (The Law of the Excluded Middle)
2. A statement cannot be both true and false. (The Law of Contradiction)

3. The converse of a true statement is not necessarily true.
4. The converse of a definition is always true.
5. For a theorem to be true, it must be true for all cases.
6. A statement is false if one false instance of the statement exists.
7. The inverse of a true statement is not necessarily true.
8. The contrapositive of a true statement is true and the contrapositive of a false statement is false.
9. If the converse of a true statement is true, then the inverse is true. Likewise, if the converse is false, the inverse is false.
10. Statements that are either both true or both false are said to be **logically equivalent**.

NECESSARY AND SUFFICIENT CONDITIONS

Let P and Q represent statements. "If P, then Q" is a conditional statement in which P is a sufficient condition for Q, and similarly Q is a necessary condition for P.

Example:

Consider the statement: "If it rains, then Jane will go to the movies." "If it rains" is a sufficient condition for Jane to go to the movies. "Jane will go to the movies" is a necessary condition for rain to have occurred.

Note that for the statement given, "If it rains" may not be the only condition for which Jane goes to the movies; however, it is a *sufficient* condition. Likewise, "Jane will go to the movies" will certainly not be the only result from a rainy weather condition (for example, "the ground will get wet" is another likely conclusion). However, knowing that Jane went to the movies is a *necessary* condition for rain to have occurred.

In the biconditional statement "P if and only if Q," P is a necessary and sufficient condition for Q, and vice versa.

Example:

Consider the statement "Rick gets paid if and only if he works." "Rick gets paid" is both a sufficient and necessary condition for him to work. Also, Rick's working is a sufficient and necessary condition for him to get paid.

Thus, we have the following basic principles to add to our list of ten from the preceding section:

11. If a given statement and its converse are both true, then the conditions in the hypothesis of the statement are both necessary and sufficient for the conclusion of the statement.
12. If a given statement is true but its converse is false, then the conditions are sufficient but not necessary for the conclusion of the statement.
13. If a given statement and its converse are both false, then the conditions are neither sufficient nor necessary for the statement's conclusion.

DEDUCTIVE REASONING

An arrangement of statements that would allow you to deduce the third one from the preceding two is called a **syllogism**. A syllogism has three parts:

1. The first part is a general statement concerning a whole group. This is called the **major premise**.
2. The second part is a specific statement which indicates that a certain individual is a member of that group. This is called the **minor premise**.
3. The last part of a syllogism is a statement to the effect that the general statement which applies to the group also applies to the individual. This third statement of a syllogism is called a **deduction**.

Example:

This is an example of a properly deduced argument.

A. Major Premise: All birds have feathers.

B. Minor Premise: An eagle is a bird.

C. Deduction: An eagle has feathers.

The technique of employing a syllogism to arrive at a conclusion is called **deductive reasoning**.

If a major premise that is true is followed by an appropriate minor premise that is true, a conclusion can be deduced that must be true, and the reasoning is valid. However, if a major premise that is true is followed by an *inappropriate* minor premise that is also true, a conclusion cannot be deduced.

Example:

This is an example of an improperly deduced argument.

A. Major Premise: All people who vote are at least 18 years old.

B. Improper Minor Premise: Jane is at least 18.

C. Illogical Deduction: Jane votes.

The flaw in this example is that the major premise in statement A makes a condition on people who vote, not on a person's age. If statements B and C are interchanged, the resulting three-part deduction would be logical.

In the following we will use capital letters X, Y, Z, ... to represent sentences, and develop algebraic tools to represent new sentences formed by linking them with the above connectives. Our connectives may be regarded as operations transforming one or more sentences into a new sentence. To describe them in greater detail, we introduce symbols to represent them. You will find that different symbols representing the same idea may appear in different references.

TRUTH TABLES AND BASIC LOGICAL OPERATIONS

The **truth table** for a sentence X is the exhaustive list of possible logical values of X.

The **logical value** of a sentence X is true (or T) if X is true, and false (or F) if X is false.

NEGATION

If X is a sentence, then $\sim X$ represents the **negation**, the opposite, or the contradiction of X. Thus, the logical values of $\sim X$ are as shown in Table 8-1, where $\sim$ is called the **negation operation** on sentences.

Table 8-1 Truth Table for Negation

X	$\sim X$
T	F
F	T

Example:

For X = "Jane is eating an apple," we have

$\sim X$ = "Jane is *not* eating an apple."

The negation operation is called *unary*, transforming a sentence into a unique image sentence.

IFF

We use the symbol **IFF** to represent the expression "if and only if."

AND

For sentences X and Y, the conjunction "X AND Y," represented by $X \wedge Y$, is the sentence that is true IFF both X and Y are true. The truth table for $\wedge$ (or AND) is shown in Table 8-2, where $\wedge$ is called the **conjunction operator**.

Table 8-2 Truth Table for AND

X	Y	$X \wedge Y$
T	T	T
T	F	F
F	T	F
F	F	F

The conjunction $\wedge$ is a *binary* operation, transforming a pair of sentences into a unique image sentence.

Example:

For X = "Jane is eating an apple" and Y = "All apples are sweet," we have $X \wedge Y$ = "Jane is eating an apple AND all apples are sweet."

AND/OR

For sentences X and Y, the disjunction "X AND/OR Y," represented by $X \vee Y$, denotes the sentence that is true if either or both X and Y are true. The truth table for $\vee$ is shown in Table 8-3, where $\vee$ is called the **disjunction operator**.

Table 8-3 Truth Table for AND/OR

X	Y	$X \vee Y$
T	T	T
T	F	T
F	T	T
F	F	F

As with the conjunction operator, the disjunction is a *binary* operation, transforming the pair of sentences X, Y into a unique image sentence $X \vee Y$.

Example:

For X = "Jane is eating the apple" and Y = "Marvin is running," we have $X \vee Y$ = "Jane is eating the apple AND/OR Marvin is running."

IF-THEN

For sentences X and Y, the **implication** $X \rightarrow Y$ represents the statement "IF X THEN Y." $X \rightarrow Y$ is false IFF X is true and Y is false; otherwise, it is true. The truth table for $\rightarrow$ is shown in Table 8-4. $\rightarrow$ is referred to as the **implication operator**.

Table 8-4 Truth Table for IF-THEN

X	Y	$X \rightarrow Y$
T	T	T
T	F	F
F	T	T
F	F	T

Implication is a *binary* operation, transforming the pair of sentences X and Y into a unique image sentence $X \rightarrow Y$.

LOGICAL EQUIVALENCE

For sentences X and Y, the **logical equivalence** $X \leftrightarrow Y$ is true IFF X and Y have the same truth value; otherwise, it is false. The truth table for $\leftrightarrow$ is shown by Table 8-5, where $\leftrightarrow$ represents logical equivalence, "IFF."

Table 8-5 Truth Table for Equivalence

X	Y	$X \leftrightarrow Y$
T	T	T
T	F	F
F	T	F
F	F	T

Example:

For X = "Jane eats apples" and Y = "apples are sweet," we have $X \leftrightarrow Y$ = "Jane eats apples IFF apples are sweet."

Equivalence is a *binary* operation, transforming pairs of sentences X and Y into a unique image sentence $X \leftrightarrow Y$. The two sentences X, Y for which $X \leftrightarrow Y$ are said to be logically equivalent.

LOGICAL EQUIVALENCE VERSUS "MEANING THE SAME"

Logical equivalence ($\leftrightarrow$) is not the same as an equivalence of meanings. Thus, if Jane is eating an apple and Barbara is frightened of mice, then for $X =$ "Jane is eating an apple" and $Y =$ "Barbara is frightened of mice," X and Y are logically equivalent, since both are correct. However, they do not have the same meaning. Statements having the same meaning are, for example, the double negative $\sim\sim X$ (not-not) and X itself.

THEOREM 1—Double Negation Equals Identity

For any sentence X,

$\sim\sim X \leftrightarrow X$.

FUNDAMENTAL PROPERTIES OF OPERATIONS

THEOREM 2—Properties of Conjunction Operation

For any sentences X, Y, Z, the following properties hold:

1. Commutativity: $X \wedge Y \leftrightarrow Y \wedge X$
2. Associativity: $X \wedge (Y \wedge Z) \leftrightarrow (X \wedge Y) \wedge Z$

THEOREM 3—Properties of Disjunction Operation

For any sentences X, Y, Z, the following properties hold:

1. Commutativity: $X \vee Y \leftrightarrow Y \vee X$
2. Associativity: $X \vee (Y \vee Z) \leftrightarrow (X \vee Y) \vee Z$

THEOREM 4—Distributive Laws

For any sentences X, Y, Z, the following laws hold:

1. $X \vee (Y \wedge Z) \leftrightarrow (X \vee Y) \wedge (X \vee Z)$
2. $X \wedge (Y \vee Z) \leftrightarrow (X \wedge Y) \vee (X \wedge Z)$

THEOREM 5—De Morgan's Laws for Sentences

For any sentences X, Y, the following laws hold:

1. $\sim(X \wedge Y) \leftrightarrow (\sim X) \vee (\sim Y)$
2. $\sim(X \vee Y) \leftrightarrow (\sim X) \wedge (\sim Y)$

Proof of Part 1 of Theorem 5

We can prove $\sim(X \wedge Y) \leftrightarrow (\sim X) \vee (\sim Y)$ by developing a truth table over all possible combinations of X and Y and observing that all values assumed by the sentences are the same. To this end, we first evaluate the expression $\sim(X \wedge Y)$ in Table 8-6a.

Table 8-6a Truth Table for Negation of Conjunction

X	Y	$X \wedge Y$	$\sim(X \wedge Y)$
T	T	T	F
T	F	F	T
F	T	F	T
F	F	F	T

Now we evaluate $(\sim X) \vee (\sim Y)$ in Table 8-6b.

Table 8-6b Truth Table for Disjunction of Negation

X	Y	$\sim X$	$\sim Y$	$(\sim X) \vee (\sim Y)$
T	T	F	F	F
T	F	F	T	T
F	T	T	F	T
F	F	T	T	T

The last columns of the truth tables coincide, proving our assertion.

THEOREM 6—Two Logical Identities

For any sentences X, Y, the sentences X and $(X \wedge Y) \vee (X \wedge \sim Y)$ are logically equivalent. That is,

$$(X \wedge Y) \vee (X \wedge \sim Y) \leftrightarrow X$$

This is proven in Table 8-7a.

Table 8-7a Truth Table for $(X \wedge Y) \vee (X \wedge \sim Y) \leftrightarrow X$

X	Y	$\sim Y$	$X \wedge Y$	$X \wedge \sim Y$	$(X \wedge Y) \vee (X \wedge \sim Y)$
T	T	F	T	F	T
T	F	T	F	T	T
F	T	F	F	F	F
F	F	T	F	F	F

For any sentences X, Y, the sentences X and $X \vee (Y \wedge \sim Y)$ are logically equivalent. That is

$$X \vee (Y \wedge \sim Y) \leftrightarrow X$$

This is proven in Table 8-7b.

Table 8-7b Truth Table for $X \vee (Y \wedge \sim Y) \leftrightarrow X$

X	Y	$\sim Y$	$Y \wedge \sim Y$	$X \vee (Y \wedge \sim Y)$
T	T	F	F	T
T	F	T	F	T
F	T	F	F	F
F	F	T	F	F

For any sentences X, Y, $(X \rightarrow Y)$ and $(\sim X \vee Y)$ are logically equivalent.

This is proven in Table 8-7c.

Table 8-7c

X	Y	$X \rightarrow Y$	$\sim X$	$\sim X \vee Y$
T	T	T	F	T
T	F	F	F	F
F	T	T	T	T
F	F	T	T	T

THEOREM 7—Proof by Contradiction

For any sentences X, Y, the following holds:

$$X \rightarrow Y \leftrightarrow \sim Y \rightarrow \sim X$$

To prove this, we consider Table 8-8.

Table 8-8 Truth Table for Proof by Contradiction

X	Y	$X \rightarrow Y$	$\sim Y$	$\sim X$	$\sim Y \rightarrow \sim X$
T	T	T	F	F	T
T	F	F	T	F	F
F	T	T	F	T	T
F	F	T	T	T	T

SENTENCES, LITERALS, AND FUNDAMENTAL CONJUNCTIONS

We have seen that logically equivalent sentences may be expressed in different ways, the simplest examples being that a sentence is equal to its double negation,

$$\sim\sim X \leftrightarrow X,$$

and by De Morgan's theorem,

$$X \vee Y \leftrightarrow \sim(\sim X \wedge \sim Y)$$

The significance of sentential calculus and the algebra of logic is that it provides us with a method of producing a "standard" form for representing a statement in terms of the literals. This is indeed unique and, although usually the simplest representation, it does serve as a standard form for comparison and evaluation of sentences.

SETS

You have seen the topics of set theory in most of the chapters of this book; in fact, you use set theory in many of your everyday activities. But since it is not labeled as "set theory" in most cases, you are unaware that set theory is the basis for most of your mathematical and logical thought. In this section you'll find the set theory vocabulary you should know as well as such topics as Venn diagrams for the union and intersection of sets (used in logic), laws of set operations (similar to those for operations on the real number system), and Cartesian products (used in graphs of linear functions). Let's set the stage for sets.

A **set** is defined as a collection of items. Each individual item belonging to a set is called an **element** or **member** of that set.

Sets are usually represented by capital letters, and elements by lowercase letters. If an item k belongs to a set A, we write $k \in A$ ("k is an element of A"). If k is not in A, we write $k \notin A$ ("k is not an element of A").

The order of the elements in a set does not matter:

$$\{1, 2, 3\} = \{3, 2, 1\} = \{1, 3, 2\}, \text{ etc.}$$

A set can be described in two ways:

1. element by element.
2. a rule characterizing the elements.

For example, given the set A of the whole numbers starting with 1 and ending with 9, we can describe it either as $A = \{1, 2, 3, 4, 5, 6, 7, 8, 9\}$ or as $A = \{\text{whole numbers greater than 0 and less than 10}\}$. In both methods, the description is enclosed in brackets. A kind of shorthand is often used for the second method of set description, so instead of writing out a complete sentence between the brackets, we can write instead

$$\text{A} = \{k \mid 0 < k < 10, k \text{ a whole number}\}$$

This is read as "the set of all elements k such that k is greater than 0 and less than 10, where k is a whole number."

A set not containing any members is called the **empty** or **null** set. It is written either as ϕ or $\{\ \}$.

A set is **finite** if the number of its elements can be counted.

Example:

$\{2, 3, 4, 5\}$ is finite since it has four elements.

Example:

$\{3, 6, 9, 12, ..., 300\}$ is finite since it has 100 elements.

Note: The empty set, denoted by ϕ, is finite since we can count the number of elements it has, namely zero.

Any set that is not finite is called **infinite**.

Example:

$\{1, 2, 3, 4, ...\}$

Example:

$\{..., -7, -6, -5, -4\}$

Example:

$\{x \mid x \text{ is a real number between 4 and 5}\}$

SUBSETS

> Given two sets A and B, A is said to be a **subset** of B if every member of set A is also a member of set B.

A is a *proper* subset of B if B contains at least one element not in A. We write $A \subseteq B$ if A is a subset of B, and $A \subset B$ if A is a proper subset of B.

Two sets are **equal** if they have exactly the same elements; in addition, if $A = B$, then $A \subseteq B$ and $B \subseteq A$.

Example:

Let $A = \{1, 2, 3, 4, 5\}$

$B = \{1, 2\}$

$C = \{1, 4, 2, 3, 5\}$

1. A equals C, and A and C are subsets of each other, but not proper subsets.
2. $B \subseteq A, B \subseteq C, B \subset A, B \subset C$ (B is a subset of both A and C. In particular, B is a proper subset of A and C).

Two sets are **equivalent** if they have the same *number* of elements.

Example:

$O = \{3, 7, 9, 12\}$ and $E = \{4, 7, 12, 19\}$. O and E are equivalent sets, since each one has four elements.

Example:

$F = \{1, 3, 5, 7, ..., 99\}$ and $G = \{2, 4, 6, 8, ..., 100\}$. F and G are equivalent sets, since each one has 50 elements.

Note: If two sets are equal, they are automatically equivalent.

A **universal set** U is a set from which other sets draw their members. If A is a subset of U, then the complement of A, denoted A', is the set of all elements in the universal set that are not elements of A.

Example:

If $U = \{1, 2, 3, 4, 5, 6, ...\}$ and $A = \{1, 2, 3\}$, then $A' = \{4, 5, 6, ...\}$.

Figure 8-1 illustrates this concept through the use of a simple **Venn diagram**.

Figure 8-1

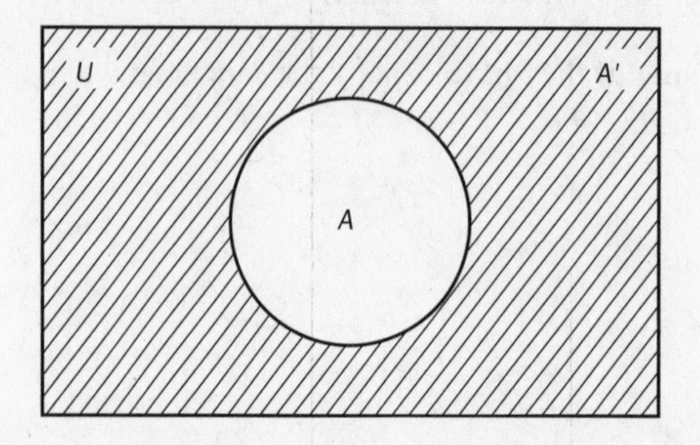

A Venn diagram is a visual way to show the relationships among or between sets that share something in common. Usually, the Venn diagram consists of two or more overlapping circles, with each circle representing a set of elements, or members. If two circles overlap, the members in the overlap belong to both sets; if three circles overlap, the members in the overlap belong to all three sets. Although Venn diagrams can be formed for any number of sets, you will probably encounter only two or three sets (circles) when working with Venn diagrams. As shown in Figure 8-1, the circles are usually drawn inside a rectangle called the universal set, which is the set of all possible members in the universe being described.

Venn diagrams are organizers. They are used to organize similarities (overlaps) and differences (non-overlaps of circles) visually, and they can pertain to any subject. For example, if the universe is all animals, Circle A may represent all animals that live in the water, and Circle B may represent all mammals. Then whales would be in the intersection of Circles A and B, but lobsters would be only in Circle A, humans would be only in Circle B, and scorpions would be in the part of the universe that was outside of Circles A and B. These relationships are shown in Figure 8-2.

Figure 8-2

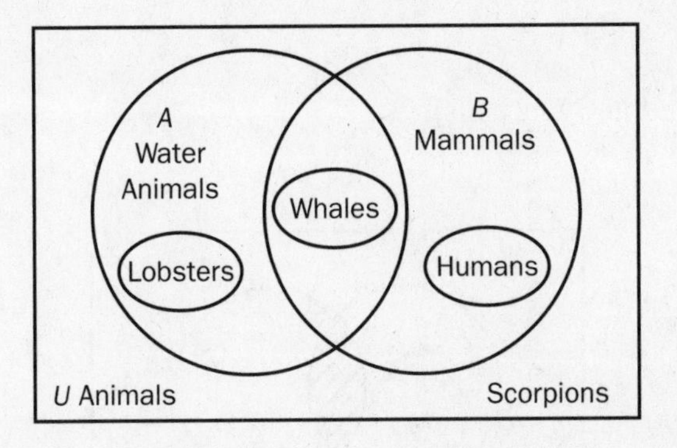

UNION AND INTERSECTION OF SETS

The **union** of two sets A and B, denoted $A \cup B$, is the set of all elements that are either in A or B or both.

Figure 8-3 is a Venn diagram for $A < B$. The shaded area represents the given operation.

Figure 8-3

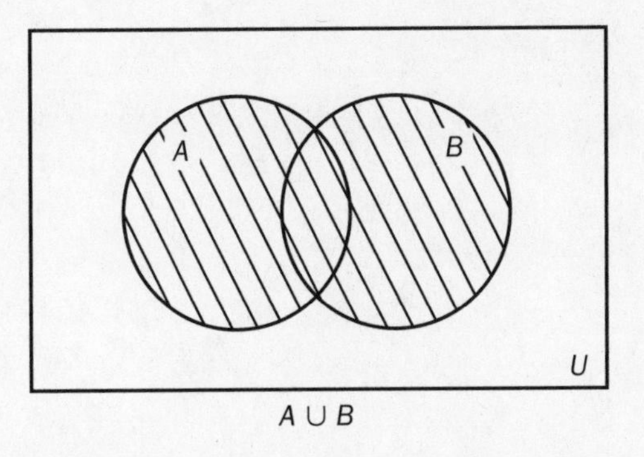

$A \cup B$

The **intersection** of two sets A and B, denoted $A \cap B$, is the set of all elements that belong to both A and B.

Figure 8-4 is a Venn diagram for $A \cap B$. The shaded area represents the given operation.

If $A = \{1, 2, 3, 4, 5\}$ and $B = \{2, 3, 4, 5, 6\}$, then $A \cup B = \{1, 2, 3, 4, 5, 6\}$ and $A \cap B = \{2, 3, 4, 5\}$.

If $A \cap B = \phi$, A and B are **disjoint**.

Figure 8-4

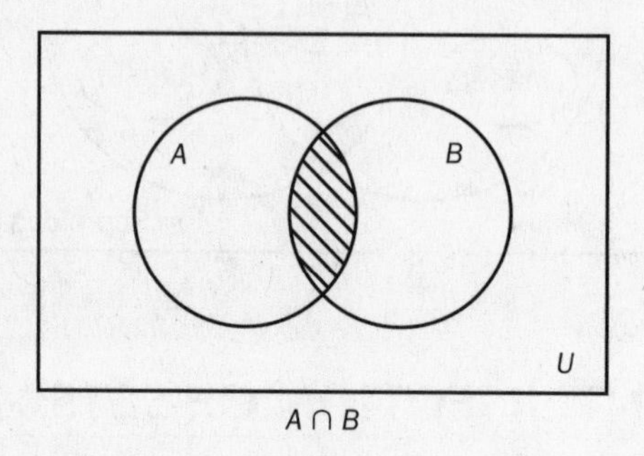

LAWS OF SET OPERATIONS

If U is the universal set, and A, B, C are any subsets of U, then the following hold for union, intersection, and complement:

Identity Laws

1a. $A \cup \phi = A$

1b. $A \cap \phi = \phi$

2a. $A \cup U = U$

2b. $A \cap U = A$

Idempotent Laws

3a. $A \cup A = A$

3b. $A \cap A = A$

Complement Laws

4a. $A \cup A' = U$

4b. $A \cap A' = \phi$

5a. $\phi' = U$

5b. $U' = \phi$

Commutative Laws

6a. $A \cup B = B \cup A$

6b. $A \cap B = B \cap A$

Associative Laws

7a. $(A \cup B) \cup C = A \cup (B \cup C)$

7b. $(A \cap B) \cap C = A \cap (B \cap C)$

Figures 8-5 and 8-6 illustrate the associative law for intersections. In Figure 8-5, the intersection of A and B is done first, and then the intersection of this result with C. In Figure 8-6, the intersection of B and C is done first, and then the intersection of this result with A. In both cases, the end result (double-hatched region) is the same.

Figure 8-5

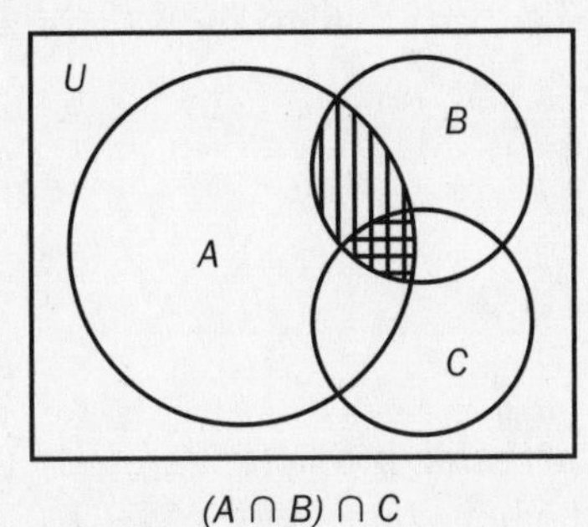

$(A \cap B) \cap C$

Figure 8-6

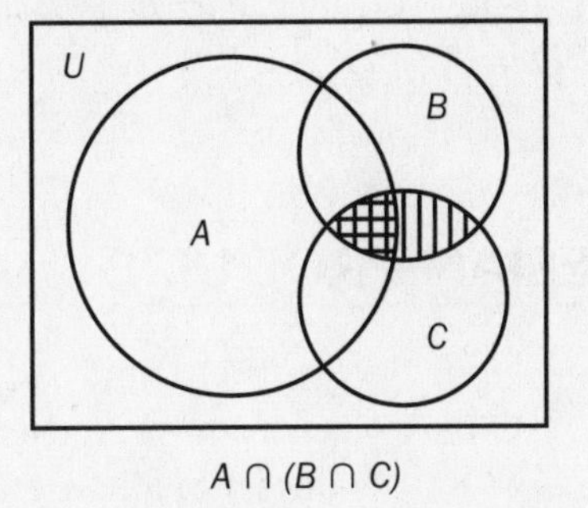

$A \cap (B \cap C)$

Distributive Laws

8a. $A \cup (B \cap C) = (A \cup B) \cap (A \cup C)$

8b. $A \cap (B \cup C) = (A \cap B) \cup (A \cap C)$

De Morgan's Laws

9a. $(A \cup B)' = A' \cap B'$

9b. $(A \cap B)' = A' \cup B'$

> The **difference** of two sets, A and B, written as $A - B$, is the set of all elements that belong to A but do not belong to B.

Example:

$J = \{10, 12, 14, 16\}, K = \{9, 10, 11, 12, 13\}$

$J - K = \{14, 16\}$. Note that $K - J = \{9, 11, 13\}$.

In general, $J - K \neq K - J$.

Example:

$T = \{a, b, c\}, V = \{a, b, c, d, e, f\}$.

$T - V = \phi$, whereas $V - T = \{d, e, f\}$.

Note, in this example, that T is a proper subset of V. In general, whenever set A is a proper subset of set B, $A - B = \phi$.

If set P is any set, then $P - \phi = P$ and $\phi - P = \phi$. Also if P and Q are any sets, $P - Q = P \cap Q'$.

CARTESIAN PRODUCT

Given two sets M and N, the **Cartesian product**, denoted $M \times N$, is the set of all ordered pairs of elements in which the first component is a member of M and the second component is a member of N.

Often, the elements of the Cartesian product can be found by making a table with the elements of the first set as row headings and the elements of the second set as column headings, and the elements of the table the pairs formed from these elements.

Example:

$M = \{1, 3, 5\}, N = \{2, 8\}$
The Cartesian product $M \times N = \{(1, 2), (1, 8), (3, 2), (3, 8), (5, 2), (5, 8)\}$
We can easily see that these are all of the elements of $M \times N$ and the only elements of $M \times N$ by looking at Table 8-9:

Table 8-9 ***M* × *N***

	2	8
1	1, 2	1, 8
3	3, 2	3,8
5	5, 2	5, 8

Example:

$W = \{a, b, c\}$, $Y = \{a, g, h\}$
The Cartesian product $W \times Y = \{(a, a), (a, g), (a, h), (b, a), (b, g), (b, h), (c, a), (c, g), (c, h)\}$

The elements for this Cartesian product are shown in Table 8-10.

Table 8-10 ***W* × *Y***

	a	g	h
a	a, a	a, g	a, h
b	b, a	b, g	b, h
c	c, a	c, g	c, h

In the first example above, since M has 3 elements and N has 2 elements, $M \times N$ has $3 \times 2 = 6$ elements. In the second example above, since W has 3 elements and Y has 3 elements, $W \times Y$ has $3 \times 3 = 9$ elements. In general, if the first set has x elements and the second set has y elements, the Cartesian product will have xy elements.

Drill Questions

1. If P and Q represent statements, which one of the following is equivalent to "Not P and not Q"?

   (A) Not P or not Q
   (B) Not P or Q
   (C) Not (P or Q)
   (D) Not (P and Q)

2. Let R and S represent statements. Consider the following:

   I. If R then S
   II. Not R and S
   III. If S then R

   Which of the above statements is (are) equivalent to the statement "R is a necessary condition for S"?

   (A) Only I
   (B) I and II
   (C) II and III
   (D) Only III

3. What is the inverse of the statement "If it is snowing, then people stay indoors"?

   (A) If people stay indoors, then it is snowing.
   (B) If it is not snowing, then people do not stay indoors.
   (C) If people do not stay indoors, then it is not snowing.
   (D) If it is not snowing, then people stay indoors.

4. What is the negation for the statement "Image is important or personality matters"?

   (A) Image is important and personality does not matter.
   (B) Image is not important or personality does not matter.
   (C) Image is important or personality does not matter.
   (D) Image is not important and personality does not matter.

5. Given any two statements P and Q, where Q is a false statement, which one of the following *must* be false?

   (A) Not P and Q
   (B) Not (P and Q)
   (C) P or not Q
   (D) P implies Q

6. "All of P is in Q and some of R is in P." Based on the previous statement, which one of the following is a valid conclusion?

   (A) Some of R is not in Q.
   (B) All of R is in Q.
   (C) Some of R is in Q.
   (D) None of R is in Q.

7. Given three statements, P, Q, and R, suppose it is known that R is true. Which one of the following must be true?

   (A) $(P \wedge Q) \rightarrow R$
   (B) $R \rightarrow (P \vee Q)$
   (C) $(P \rightarrow Q) \wedge R$
   (D) $(R \rightarrow Q) \vee P$

8. Suppose that set K has 12 elements and set L has 3 elements. How many elements are there in the Cartesian product $K \times L$?

   (A) 4
   (B) 9
   (C) 15
   (D) 36

9. If $A = \{x \mid x \text{ is an even integer less than } 10\}$ and $B = \{\text{all negative numbers}\}$, which one of the following describes $A \cap B$?

   (A) {all negative numbers and all positive even integers}
   (B) {all negative numbers}
   (C) {all negative even integers}
   (D) {all negative odd integers}

10. Consider the Venn diagram shown below. Which one of the following correctly describes the shading?

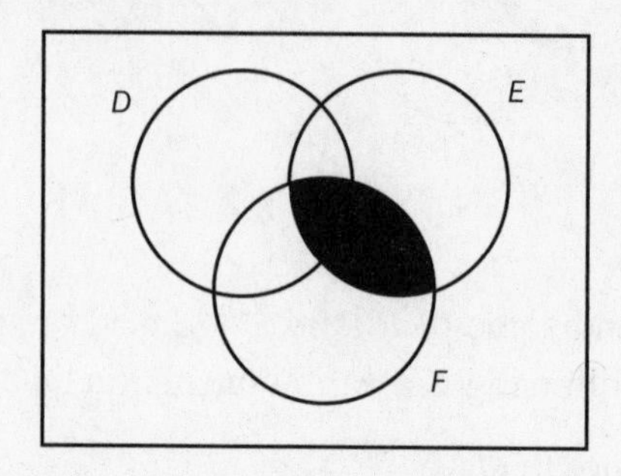

(A) $E \cap F$
(B) $D \cap E \cap F$
(C) $D \cap E$
(D) $D \cap F$

11. Which one of the following is an example of disjoint sets?

(A) {0, 1, 2, 3} and {3, 2, 1, 0}
(B) {0, 2, 4, 6} and {2, 4, 6, 8}
(C) {0, 3, 6, 9} and {9, 16, 25, 36}
(D) {0, 4, 8, 12} and {6, 10, 14, 18}

12. If the universal set $U = \{x \mid x \text{ is a positive odd integer less than } 30\}$, $R = \{1, 5, 7\}$, and $S = \{1, 3, 7, 11, 13\}$, how many elements are in $(R \cap S)'$?

(A) 15
(B) 13
(C) 7
(D) 2

13. If $P \subseteq Q$, which one of the following conclusions must be true?

(A) P is either equal to Q or P is a proper subset of Q.
(B) P is a proper subset of Q.
(C) Q is a proper subset of P.
(D) P is either equal to Q or P is the empty set.

14. Given any two sets F and G, which one of the following is not necessarily true?

    (A) $F \cup G = G \cup F$
    (B) $F \cap G = G \cap F$
    (C) $F - G = G - F$
    (D) $F \cap F' = \phi$

15. Consider the Venn diagram shown below.

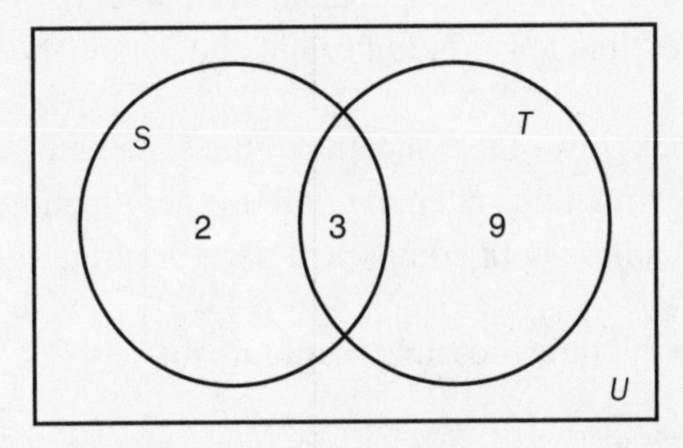

If the given numbers represent the number of elements in each region, how many elements are in $T - S$?

(A) 10
(B) 9
(C) 7
(D) 6

Answers to Drill Questions

1. **(C)** One of De Morgan's Laws for sentences is $\sim(X \vee Y) \leftrightarrow - X \wedge \sim Y$. Substituting P for X and Q for Y, we have $\sim(P \vee Q) \leftrightarrow \sim P \wedge \sim Q$. This statement is read as follows: "Not (P or Q)" is equivalent to "Not P and not Q."

2. **(D)** If R is a necessary condition for S, then by definition S implies R. Also, the statement that "S implies R" is equivalent to the statement "If S then R," which represents item III. Neither of items I or II is equivalent to the given statement.

3. **(B)** The inverse of "If P then Q" is "If not P then not Q." Let P represent the statement "It is snowing" and let Q represent the statement "People stay

indoors." By substitution, for the statement "If it is snowing, then people stay indoors," the inverse statement is "If it is not snowing, then people do not stay indoors."

4. **(D)** The negation of $(P \vee Q)$ is $\sim(P \vee Q)$. By one of De Morgan's Laws, the statement $\sim(P \vee Q)$ is equivalent to $\sim P \wedge \sim Q$. Let P represent the statement "Image is important" and let Q represent the statement "Personality matters." Then the negation for "Image is important or personality matters" is "Not (Image is important or personality matters)." This latter statement is equivalent to "Image is not important and personality does not matter."

5. **(A)** Given that Q is a false statement, the statement "Not P and Q" must be false. Any compound statement with the conjunction operator (which is "and") is false unless both component parts are true.

6. **(C)** Here are the three possible diagrams for sets P, Q, and R.

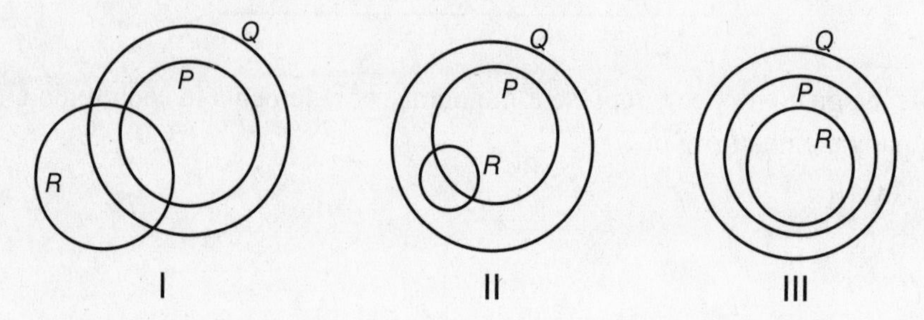

The statement "Some of R is in Q" is true for each of diagrams I, II, and III. Choice (A) is incorrect because it is false for diagrams II and III. Choice (B) is incorrect because it is false for diagram I. Choice (D) is incorrect because it is false for all three diagrams. Note that the statement "Some of R is in P" is true even if all of R is actually in P. The word "some" means "at least one."

7. **(A)** A conditional statement "$X \rightarrow Y$" is true in all instances except when X is true and Y is false. Let X be represented by $P \wedge Q$ and let Y be represented by R. Since R is known to be true and we do not know the truth value of "$P \wedge Q$," we have either "True $\rightarrow$ True" or "False $\rightarrow$ True." In either case "$(P \wedge Q) \rightarrow R$" must be true.

8. **(D)** Given two sets K and L, the Cartesian product $K \times L$ is the set consisting of all ordered pairs in which the first element is chosen from K and the

second element is chosen from L. For example, if set K contains the element a and set L contains the element b, then one of the elements of $K \times L$ is (a, b). Since we know that K has 12 elements and L has 3 elements, we conclude that there are $(12)(3) = 36$ different ordered pairs for $K \times L$.

9. **(C)** In roster form, $A = \{..., -8, -6, -4, -2, 0, 2, 4, 6, 8\}$. Although we cannot write set B in roster form, we can determine that the elements common to both A and B can be represented as $\{..., -8, -6, -4, -2\}$, which is the set of all negative even integers.

10. **(A)** The shaded region represents all elements common to both sets E and F. This is the definition of the intersection of sets E and F, written as $E \cap F$. Note that the presence of set D becomes incidental, and does not affect the correct answer choice.

11. **(D)** Disjoint sets are those that do not contain any common elements, such as $\{0, 4, 8, 12\}$ and $\{6, 10, 14, 18\}$.

12. **(B)** Written in roster form, $U = \{1, 3, 5, 7, ..., 29\}$, which contains 15 elements. $R \cap S$ represents the elements common to both R and S, so $R \cap S = \{1, 7\}$. Finally, $(R \cap S)'$ represents the elements in U that do not belong to $R \cap S$. Thus, $(R \cap S)'$ must contain $15 - 2 = 13$ elements.

13. **(A)** $P \subseteq Q$ means that each element of set P is also an element of set Q. This implies that either (a) P and Q are identical, or (b) Q contains all the elements of P, plus at least one element not found in P. By definition, if part (b) applies, then P is a proper subset of Q.

14. **(C)** $F - G$ represents the set of elements in F that do not belong to G, whereas $G - F$ represents the set of elements in G that do not belong to F. Unless these sets are equivalent, $F - G \neq G - F$. For example, let $F = \{1, 3, 6\}$ and let $G = \{1, 5, 8, 9\}$. Then $F - G = \{3, 6\}$, but $G - F = \{5, 8, 9\}$.

15. **(B)** $T - S$ is the set of elements in T that do not belong to S. According to the given Venn Diagram, there are 9 elements that fit this description.

PRACTICE TEST 1

CLEP College Mathematics

Also available at the REA Study Center (*www.rea.com/studycenter*)

This practice test is also available online at the REA Study Center. The CLEP College Mathematics test is only offered as a computer-based exam; therefore, we recommend that you take the online version of the practice test to receive these added benefits:

- **Timed testing conditions** – Gauge how much time you can spend on each question.
- **Automatic scoring** – Find out how you did on the test, instantly.
- **On-screen detailed explanations of answers** – Learn not just the correct answer, but also why the other answer choices are incorrect.
- **Diagnostic score reports** – Pinpoint where you're strongest and where you need to focus your study.

PRACTICE TEST 1

CLEP College Mathematics

(Answer sheets appear in the back of the book.)

TIME: 90 Minutes
60 Questions

Directions: An online scientific calculator will be available for the questions in this test.

Some questions will require you to select from among four choices. For these questions, select the BEST of the choices given.

Some questions will require you to type a numerical answer in the box provided.

Notes: (1) Unless otherwise specified, the domain of any function f is assumed to be the set of all real numbers x for which $f(x)$ is a real number.

(2) i will be used to denote $\sqrt{-1}$

(3) Figures that accompany questions are intended to provide information useful in answering the questions. All figures lie in a plane unless otherwise indicated. The figures are drawn as accurately as possible EXCEPT when it is stated in a specific question that the figure is not drawn to scale.

1. Which one of the following is equivalent to the negation of the statement "Cats are friendly and Bob has a hamster"?

   (A) If cats are friendly, then Bob does not have a hamster.
   (B) If Bob has a hamster, then cats are friendly.
   (C) If cats are not friendly, then Bob has a hamster.
   (D) If Bob does not have a hamster, then cats are not friendly.

2. 50 miles per hour is the same as

   I. 73.33 ft/sec
   II. 1,466.67 yards/min

   (A) I only
   (B) II only
   (C) Both I and II
   (D) Neither I nor II

3. If x is an odd integer and y is even, then which of the following must be an even integer?

   I. $2x + 3y$
   II. xy
   III. $x + y - 1$

   (A) I only
   (B) II only
   (C) I, II, and III
   (D) II and III only

4. An ordinary six-sided cube, with its sides numbered 1 through 6, is rolled twice. The probability of rolling any of the six numbers is equally likely. What is the probability that on two consecutive rolls of the cube, a number less than 3 appears on the first roll and the number 5 appears on the second roll?

5. Not counting the empty set, how many proper subsets are there for R = {2, 3, 4}?

(A) 5
(B) 6
(C) 7
(D) 8

6. The average size of a house in 2010 was 2,463 ft^2. This is 145% of the average size of a house in 1980. What was the average size of a house in 1980?

(A) 1,108 ft^2
(B) 1,589 ft^2
(C) 1,699 ft^2
(D) 3,572 ft^2

7. Let P, Q, and R represent statements where P is true, Q is false, and R is false. Which one of the following is a true statement?

(A) (P and R) or Q
(B) (P implies Q) and Not R
(C) Not P or (Q and R)
(D) Not P implies (Q and R)

8. Given that $i = \sqrt{-1}$, what is the simplified expression for $3i^3 - 4i^2 + 5i$?

(A) $-2i - 4$
(B) $-2i + 4$
(C) $2i - 4$
(D) $2i + 4$

9. The federal debt is the amount the government owes after borrowing the money it needs to pay for its expenses. It measures how much of government spending is financed by debt rather than taxation. If the federal debt was 3.21 trillion dollars in 1990 and 13.79 trillion dollars in 2010, find the approximate percentage increase in the federal debt from 1990 to 2010.

   (A) 3%
   (B) 23%
   (C) 77%
   (D) 330%

10. Payday lending has been legalized in some states. It involves lending money, usually at a very high interest rate, until a borrower can pay it off with his paycheck. If a borrower secures a payday loan of $625 for 15 days at 450% APR simple interest, to the nearest dollar, how much must he pay back?

    (A) $117
    (B) $636
    (C) $742
    (D) $906

11. A highway study of 15,000 vehicles that passed by a checkpoint found that their speeds were normally distributed with a mean of 59 mph and a standard deviation of 6 mph. How many of the vehicles had speeds greater than 65 mph?

    (A) 375
    (B) 2,400
    (C) 5,100
    (D) 9,900

12. You win $2 million in the lottery. You invest half that money in Bank M at 3.02% interested compounded daily while the other half is invested in Bank N at 3.05% interest compounded quarterly. What is the approximate difference between the interest earned in the two banks after the first year?

    (A) $0
    (B) $192
    (C) $264
    (D) $300

13. If $m^x \cdot m^7 = m^{28}$ and $(m^5)^y = m^{15}$, what is the value of $x + y$?

    (A) 31
    (B) 24
    (C) 14
    (D) 12

14. Which is largest?

    (A) $\log_6 4 + \log_6 9$
    (B) $\log_5 15 - \log_5 3$
    (C) $\log_4 \frac{1}{4}$
    (D) $\log 100^{1/4}$

15. Suppose $S = \{5, 6, 9\}$ and $T = \{7, 8, 9\}$. Which one of the following ordered pairs is *NOT* in the Cartesian Product of $T \times S$?

    (A) (9, 9)
    (B) (8, 5)
    (C) (6, 8)
    (D) (7, 6)

16. The future value of an investment 75 years from now is $75,000. It was invested at 5% compounded semi-annually. What is the present value?

    (A) $1,000
    (B) $1,777
    (C) $1,847
    (D) $2,102

17. A fair coin is tossed 4 times. What are the odds of getting exactly two heads?

    (A) 3:8
    (B) 8:3
    (C) 5:3
    (D) 3:5

18. Given the following list of six numbers:

$$\pi, \sqrt{5}, \sqrt{\frac{4}{25}}, -.212, 5\frac{2}{7}, \text{ and } .1\overline{8}$$

How many of these numbers are irrational?

19. The mean of Sheila's five exam scores is 78. She will be taking three more exams. Assuming that each exam is given the same weight, what must her mean score be on the remaining exams in order to attain a mean score of 84 on all 8 exams?

(A) 94
(B) 92
(C) 90
(D) 88

20. Look at the triangle below.

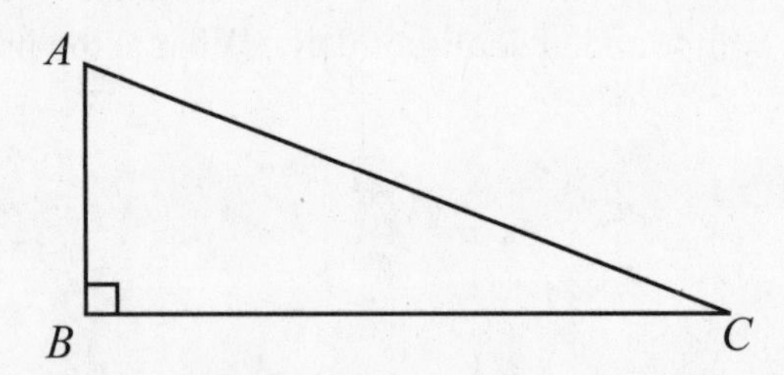

AB is perpendicular to BC. $AB = 10$ and $AC = 26$. What is the area of the triangle?

(A) 260
(B) 240
(C) 130
(D) 120

21. Which graph does NOT represent a function ($y = f(x)$)?

(A)

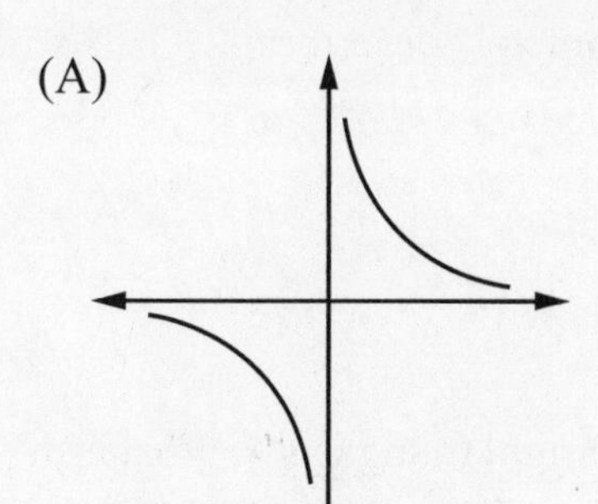

(C)

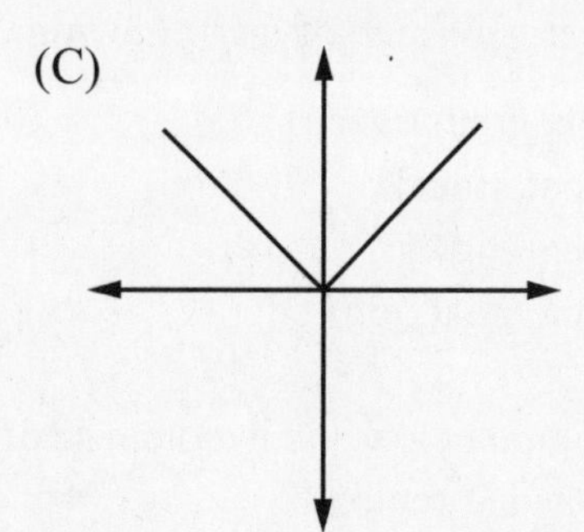

(B)

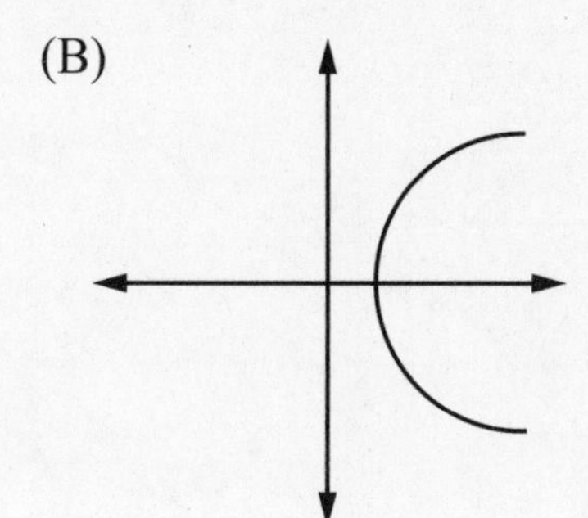

(D)

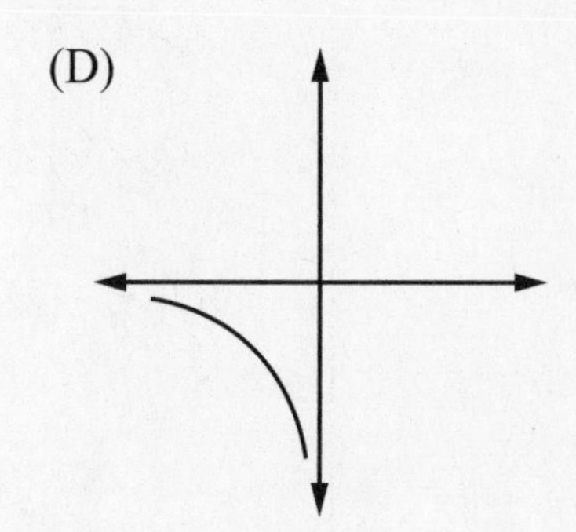

22. The cost of a men's shirt at a department store is \$15. The store marks up the price 80% and then has a sale advertising a 30% markdown. What is the percentage markup of the shirt when it is sold at the sale price?

(A) 14.4%
(B) 26%
(C) 30%
(D) 50%

23. Which one of the following has the lowest value?

(A) $|-8| - |3|$
(B) $-|-8 - 3|$
(C) $|-3 + 8| - |-8 + 3|$
(D) $-|3 - 8|$

24. Let $U = \{$cat, dog, frog, goat, horse, pig, tiger$\}$, $A = \{$dog, frog, horse, pig$\}$, and $B = \{$dog, goat, pig, tiger$\}$. Define A' as the elements in set U that are not in set A. Which of the following completely describes $A' \cap B$?

    (A) {cat, goat, tiger}
    (B) {goat, tiger}
    (C) {dog, pig}
    (D) {dog, goat, pig}

25. In the figure below, the hypotenuse of the right triangle is 4. What is the area of the shaded region?

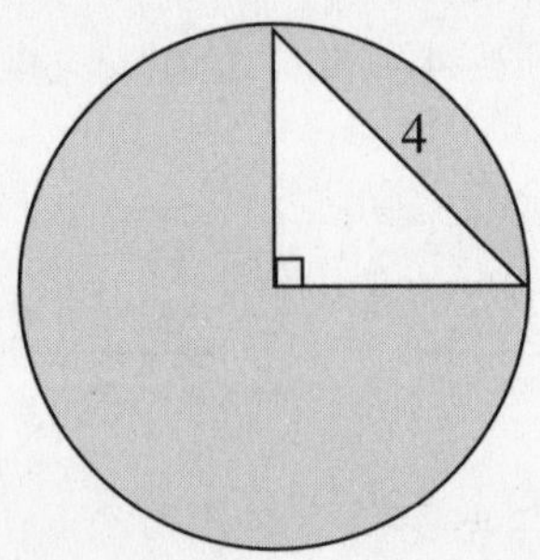

    (A) $16\pi - 8$
    (B) $16\pi - 16$
    (C) $8\pi - 4$
    (D) $8\pi - 8$

26. On a cruise ship, movies are shown at night. Popcorn and soft drinks are available. 100 people attend a showing of a movie and the number of people having popcorn and a soft drink are shown in the Venn diagram below. Classify the two events: having popcorn and having a soft drink.

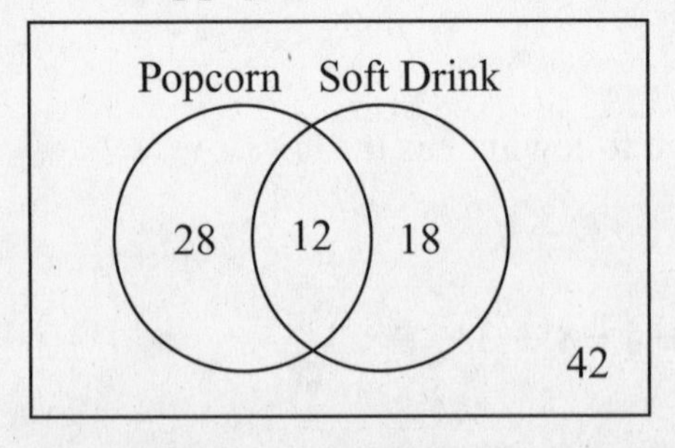

    (A) Mutually exclusive and independent
    (B) Mutually exclusive but not independent
    (C) Independent but not mutually exclusive
    (D) Neither mutually exclusive nor independent

27. If $A \subset C$ and $B \subset C$, which of the following statements is true?

(A) The set $A \cup B$ is also a subset of C.
(B) The complement of A is also a subset of C.
(C) The complement of B is also subset of C.
(D) The union of $\overline{A}$ and $\overline{B}$ is also a subset of C.

28. Which one of the following is equivalent to the statement "If Joan sings, then I will play my guitar"?

(A) Joan sings and I will play my guitar.
(B) Joan does not sing and I will not play my guitar.
(C) Joan sings or I will not play my guitar.
(D) Joan does not sing or I will play my guitar.

29. During the 1970s the number of people in the military began to decrease. The chart below shows the number of military personnel, in thousands, in each of the 4 branches of the military between 1970 and 1975. Which branch had the biggest percentage decrease?

	Army	Navy	Air Force	Marines
1975	784	535	612	196
1980	777	527	578	188

(A) Amy
(B) Navy
(C) Air Force
(D) Marines

30. A floor that measures 10 feet by 20 feet is to be tiled with square tiles that are 36 square inches in area. How many tiles are needed to cover the entire floor?

31. During a sale in an appliance store, 250 people came into the store. 55 people were interested in looking at washers, 45 were interested in looking at dryers, and 25 were interested in both. What is the probability that someone was interested in either washers or dryers?

(A) 4%
(B) 10%
(C) 30%
(D) 40%

32. An analysis of commercial airplanes compares the number of seats and the rate of fuel consumption in gallons per hour as shown by the scatterplot below. The regression line is drawn. Which of the following is closest to the increase in fuel consumption for each seat added?

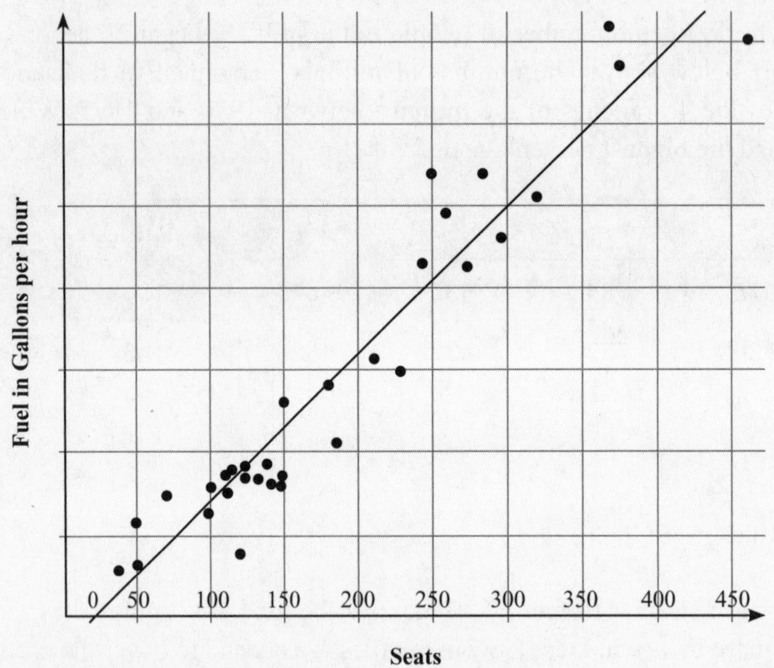

(A) 5
(B) 9
(C) 12
(D) 15

33. What is the domain of the function given by $f(x) = \dfrac{x^2 + 1}{x + 3}$?

(A) All numbers except -3
(B) All numbers except -1
(C) All numbers except 1
(D) All numbers except 3

34. In the figure below, the lines are extensions of the triangle. Find the value of $x + y - z$.

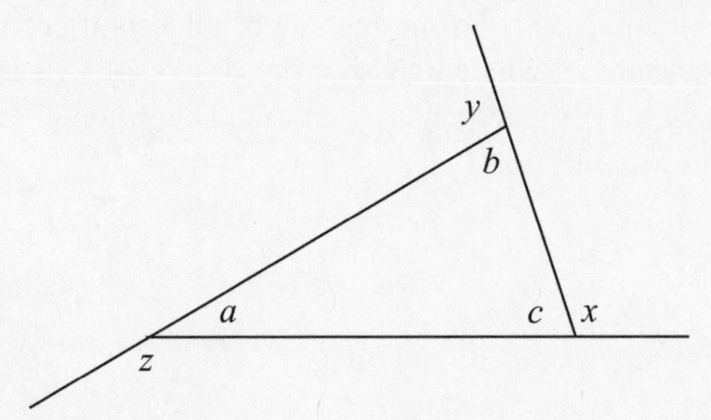

(A) $2a$
(B) $2a + 2c$
(C) 360°
(D) 720°

35. What is the median of the following data?

2, 24, 7, 10, 15, 8

(A) 7.5
(B) 8.5
(C) 9
(D) 11

36. Which one of the following is a valid argument?

(A) All rainy days are cloudy.
Yesterday was not cloudy.
Yesterday was not rainy.
(B) All trees have brown leaves.
This plant has brown leaves.
This plant is a tree.

(C) Some wolves are vicious.
This animal is vicious.
This animal is a wolf.

(D) Some people have stocks and bonds.
Charles has stocks.
Charles has bonds.

37. A nightclub has a target amount of $10 for the average amount each customer spends on drinks. They find that the average amount each customer spends is $8. So the nightclub reduces prices by 10% in the hope that customers will purchase more drinks. By what percent must customers increase their spending with the new pricing structure for the nightclub to reach its goal?

(A) 72%
(B) 172%
(C) 39%
(D) 139%

38. Which one of the following groups of data has exactly two modes?

(A) 1, 1, 3, 4, 4, 5, 5, 5
(B) 1, 1, 1, 2, 2, 2, 2
(C) 1, 2, 3, 3, 4, 4, 5, 5
(D) 1, 3, 3, 3, 4, 4, 4

39. If f is defined by $f(x) = \frac{5x - 8}{2}$ for each real number x, find the solution set for $f(x) > 2x$.

(A) $\{x \mid x > 6\}$
(B) $\{x \mid x > 8\}$
(C) $\{x \mid x < 8\}$
(D) $\{x \mid 6 < x < 8\}$

40. Look at the following Venn Diagram, for which a Roman numeral has been assigned to each region.

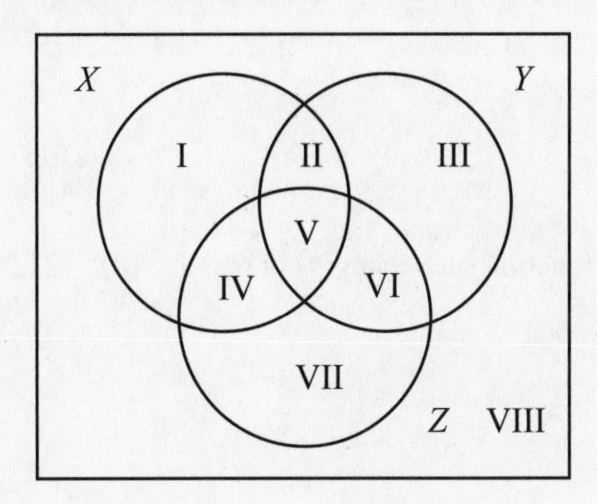

Which region(s) would include $(X \cup Y)'$?

(A) VII and VIII
(B) Only VII
(C) V, VII, and VIII
(D) Only VIII

41. The sample standard deviation, s, of a group of data is given by the formula $s = \sqrt{\dfrac{\sum_{i=1}^{n} (X_i - \overline{X})^2}{n-1}}$, where X_i represents each data, X represents the mean, and n represents the number of data. What is the sample standard deviation for the following data? 4, 5, 6, 9, 11. (Round off your answer to the nearest hundredth.)

42. For federal income tax, a person whose salary is between \$9,226 and \$37,450 pays \$923 + 15% of the amount his salary is over \$9,225. If a person makes exactly \$37,450, what percentage of his salary goes to taxes?

(A) 11.3%
(B) 13.8%
(C) 15%
(D) 17.5%

43. If g is a linear function such that $g(4) = 6$ and $g(10) = 21$, what is the value of $g(8)$?

(A) 19
(B) 18
(C) 16
(D) 10

44. A real estate broker receives a commission of 6% of the selling price of a house. If the broker received a commission of \$16,250 on the sale of the house, and the commission is taken off the price of the house at settlement, how much does the seller receive for the house?

(A) \$254,583
(B) \$270,833
(C) \$287,083
(D) \$300,433

45. A supermarket is having a sale on its store brand of soda, selling a 48-ounce bottle for \$0.99. It also sells 12-ounce cans for \$0.40 and sells its 2-liter (67.6 ounces) bottle for \$1.39. Arrange the three types from lowest to highest unit price.

(A) 2-liter – 48-ounce – can
(B) 48-ounce – 2-liter – can
(C) 48-ounce – can – 2-liter
(D) can – 48-ounce – 2-liter

46. Write $(6 \times 10^{-3})(8 \times 10^{-1})$ in standard notation.

47. A bike wheel has a radius of 12 inches. How many revolutions will it take to cover one mile? (Use 1 mile = 5,280 feet, and $\pi = \frac{22}{7}$.)

(A) 70
(B) 84
(C) 120
(D) 840

48. For which one of the following groups of data are the mean and median identical?

(A) 2, 2, 5, 6, 8, 8, 11
(B) 2, 3, 6, 6, 6, 8, 12
(C) 2, 4, 5, 6, 8, 11
(D) 2, 5, 5, 7, 8, 12

49. Which one of the following is an equation of a line containing the point $(2, -1)$ and is perpendicular to the graph of $x + 3y = 4$?

(A) $x + 3y = -1$
(B) $3x - y = 7$
(C) $3x + y = 5$
(D) $x - 3y = 5$

50. A study was done as to the height in inches of children in a school aged 9 through 16. The data is broken up by gender. The scatterplot and the two lines of best fit are shown in the figure below. What conclusion can be made from this graph?

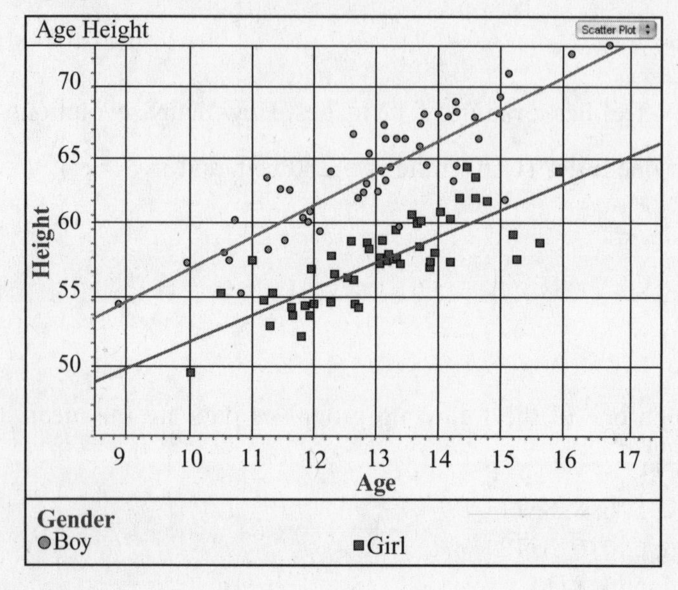

(A) There are more girls than boys
(B) There is a greater standard deviation in boy heights than girl heights
(C) Boys grow faster than girls
(D) None of these

51. The figure below is comprised of a parallelogram and a trapezoid. If the parallelogram and trapezoid have the same areas, find the value of the altitude h of the trapezoid.

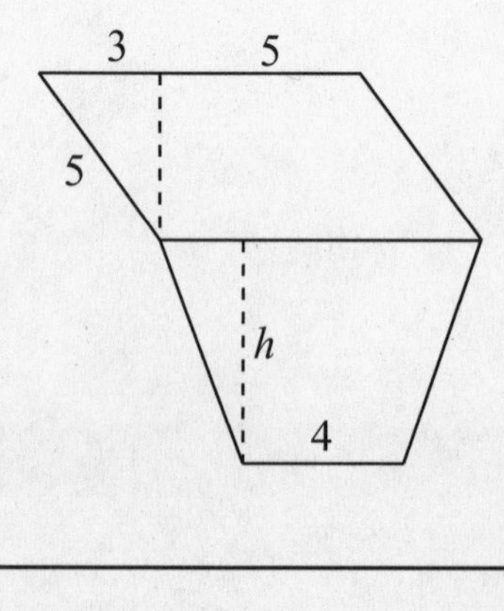

52. The function $f(x)$ is defined as follows:

$f(x) = 4x - 1$, if $x \leq -5$

$\quad\;\; = 5x + 1$, if $x > -5$

What is the value of $f(-7) + f(10)$?

53. Given a collection of nine books, in how many different ways can any four of them be placed on a shelf?

(A) 262,144
(B) 60,480
(C) 6,561
(D) 3,024

54. On January 1, 2015, Jerry receives a royalty check in the sum of $2,000 for a book he wrote. He deposits it in a bank account at 3% APR compounded daily. Six months later, he receives another royalty check for $3,000 and deposits it in the same account at the same interest rate. To the nearest dollar, how much money is in the account January 1, 2016?

(A) $5,106
(B) $5,126
(C) $5,152
(D) $5,311

55. In the figure on the next page, the graph of $f(x)$ is shown as the solid line. Which of the following is the equation of the transformation of $f(x)$ shown by the dashed line?

(A) $y = f(x + 3) - 3$
(B) $y = f(x + 3) + 3$
(C) $y = f(x - 3) - 3$
(D) $y = f(x - 3) + 3$

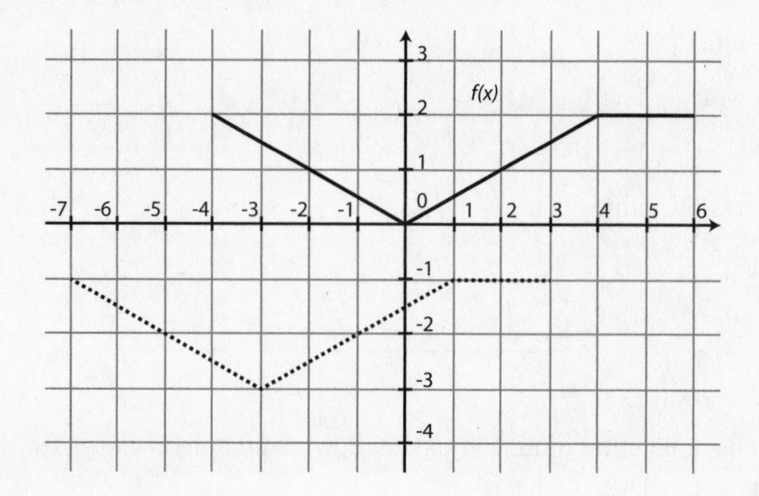

56. A charity is raffling off a 60-inch TV worth \$2,000, two 32-inch TV's worth \$500 each and five iPods worth \$230 each. A chance costs \$10 and 1,000 chances are sold. What is your mean expectation for the raffle?

    (A) winning \$4.15
    (B) losing \$5.85
    (C) winning \$2.73
    (D) losing \$7.27

57. A rectangle and a square have the same perimeter. The side of the square is 9 and the length of the rectangle is 13. What is the width of the rectangle?

58. Ed purchases a stock whose value y grows linearly as a function of time t while Noreen purchases a stock that grows exponentially according to the formula $y = 250(2)^{t+1}$ as shown by the following figure where t is measured in months. How many of the following statements are true?

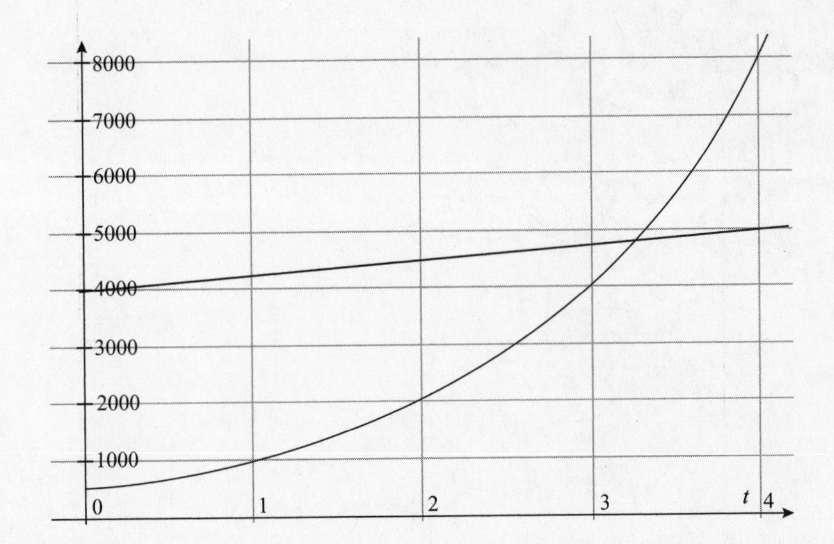

I. If they both sell their stock after 2 years, Ed's stock value will be greater.

II. When t = 0, the difference in the value of Ed's stock and Noreen's stock is $4,000.

III. When t = 4, Noreen's stock is worth 60% more than Ed's stock.

(A) 3
(B) 2
(C) 1
(D) 0

59. The number $0.\overline{8}$ is equivalent to what reduced fraction?

60. In a room of 20 people, if each person shakes hands with every other person, how many different handshakes are possible?

(A) 40
(B) 190
(C) 380
(D) 400

PRACTICE TEST 1

Answer Key

#	Answer	Chapter	#	Answer	Chapter
1.	A	8. Logic & Sets	31.	C	4. Probability & Counting
2.	C	2. Numbers	32.	B	5. Statistics & Data Analysis
3.	C	2. Numbers	33.	D	3. Algebra & Functions
4.	1/18	4. Probability & Counting	34.	A	7. Geometry
5.	B	8. Logic & Sets	35.	C	5. Statistics & Data Analysis
6.	C	6. Financial Math	36.	A	8. Logic & Sets
7.	D	8. Logic & Sets	37.	C	6. Financial Math
8.	D	3. Algebra & Functions	38.	D	5. Statistics & Data Analysis
9.	D	6. Financial Math	39.	B	3. Algebra & Functions
10.	C	6. Financial Math	40.	A	8. Logic & Sets
11.	B	5. Statistics & Data Analysis	41.	2.92	5. Statistics & Data Analysis
12.	B	6. Financial Math	42.	B	6. Financial Math
13.	B	3. Algebra & Functions	43.	C	3. Algebra & Functions
14.	A	3. Algebra & Functions	44.	A	6. Financial Math
15.	C	8. Logic & Sets	45.	A	6. Financial Math
16.	C	6. Financial Math	46.	0.0048	2. Numbers
17.	D	4. Probability & Counting	47.	D	7. Geometry
18.	2	2. Numbers	48.	A	5. Statistics & Data Analysis
19.	A	5. Statistics & Data Analysis	49.	B	3. Algebra & Functions
20.	D	7. Geometry	50.	C	5. Statistics & Data Analysis
21.	B	3. Algebra & Functions	51.	5.33	7. Geometry
22.	B	6. Financial Math	52.	22	3. Algebra & Functions
23.	B	2. Numbers	53.	D	4. Probability & Counting
24.	B	8. Logic & Sets	54.	B	6. Financial Math
25.	C	7. Geometry	55.	A	3. Algebra & Functions
26.	C	4. Probability & Counting	56.	B	4. Probability & Counting
27.	A	8. Logic & Sets	57.	5	7. Geometry
28.	D	8. Logic & Sets	58.	B	3. Algebra & Functions
29.	C	6. Financial Math	59.	8/9	2. Numbers
30.	800	7. Geometry	60.	B	4. Probability & Counting

PRACTICE TEST 1

Detailed Explanations of Answers

1. **(A)** When a statement is in the form "If P, then Q," the equivalent statement is in the form "Not P or Q." The negation of "Not P or Q" is the statement "P and not Q." Let P represent the statement "Cats are friendly." Let "not Q" represent the statement "Bob has a hamster." Then the given statement in the stem of this question is written in the form "P and not Q," so the negation will be in the form "If P, then Q." Note that Q represents the statement "Bob does not have a hamster."

2. **(C)**

I. $$\frac{50 \text{ miles}}{\text{hour}} \cdot \frac{5280 \text{ feet}}{1 \text{ mile}} \cdot \frac{1 \text{ hour}}{60 \text{ minutes}} \cdot \frac{1 \text{ minutes}}{60 \text{ seconds}} = 73.33 \frac{\text{feet}}{\text{second}}$$

II. $$\frac{50 \text{ miles}}{\text{hour}} \cdot \frac{5280 \text{ feet}}{1 \text{ mile}} \cdot \frac{1 \text{ yard}}{3 \text{ feet}} \cdot \frac{1 \text{ hour}}{60 \text{ minutes}} = 1{,}466.67 \frac{\text{yards}}{\text{minute}}$$

3. **(C)**

I. An odd integer times two will become an even integer. An even integer times any number will remain even. The sum of two even numbers is also an even number. Therefore, $2x + 3y$ must be even.

II. An even integer times any number will remain even. Therefore, xy must be even.

III. The sum of an odd integer and an even integer is odd. An odd integer minus one will become even. Therefore, $x + y - 1$ must be even.

4. The correct answer is $\frac{1}{18}$. A number less than 3 means 1 or 2. The probability of 1 or 2 appearing on the first roll is $\frac{2}{6} = \frac{1}{3}$. The probability of 5 appearing on the second roll is $\frac{1}{6}$. Since these events are independent, the

probability that both will occur is the product of these probabilities, which is $\frac{1}{3} \cdot \frac{1}{6} = \frac{1}{18}$.

5. **(B)** Given a set of n elements, the number of proper subsets is given by the expression $2^n - 1$. The set of proper subsets does not include the set itself. However, the expression $2^n - 1$ does include the empty set. Thus, the answer is $2^n - 2 = 2^3 - 2 = 6$.

6. **(C)** We want to answer the question: 145% of what number is 2,463?

$$\frac{145}{100}x = 2463 \Rightarrow 145x = 246{,}300$$

$$x = \frac{246300}{145} = 1698.62$$

The average size of a house in 1980 was 1,699 ft^2.

7. **(D)** Not P becomes false. Q and R is also false. A statement that reads "False implies false" is always true. Answer choice (A) is wrong because P and R is false, so it reads "False or False." Answer choice (B) is wrong because P implies Q is false, so it reads "False and True," which is false. Answer choice (C) is wrong because Q and R is false, so it reads "False or False."

8. **(D)**

$i^2 = -1$ and $i^3 = i^2(i) = (-1)(i) = -i$. So, $3i^3 - 4i^2 + 5i$
$= (3)(-i) - (4)(-1) + 5i = 5i - 3i + 4 = 2i + 4$.

9. **(D)** $\%\text{ increase} = \dfrac{\text{New debt} - \text{Original debt}}{\text{Original debt}} = \dfrac{13.79 - 3.21}{3.21} = 3.295 \approx 330\%$

10. **(C)** $P = 625,\ r = 450\% = 4.5,\ t = \dfrac{15\text{ days}}{360\text{ days}} = \dfrac{15}{360}$

$I = 625(4.5)\left(\dfrac{15}{360}\right) = 35$. So the simple interest due is \$117.19.

$A = P + I = 625 + 117 = \$742$

11. **(B)** 65 mph is one standard deviation above the mean. Since there is 68% of the data within one standard deviation of the mean, there is 34% of

data between the mean and 65 mph. So there is 16% of the data above one standard deviation. 0.16(15,000) = 2,400.

12. **(B)** Bank M: $A = 1000000\left(1+\dfrac{0.0302}{360}\right)^{360(1)} = \$1{,}030{,}659 \Rightarrow$

Interest = \$30,659

Bank N: $A = 1000000\left(1+\dfrac{0.0305}{4}\right)^{360(4)} = \$1{,}030{,}851 \Rightarrow$

Interest = \$30,851

Difference: 30,851 − 30,659 = \$192.

13. **(B)** For the first equation, $x + 7 = 28$, so $x = 21$. For the second equation, $5y = 15$, so $y = 3$. Then $x + y = 24$.

14. **(A)** $\log_6 4 + \log_6 9 = \log_6(4 \cdot 9) = \log_6 36 \Rightarrow 6^x = 36 \Rightarrow x = 2$

$\log_5 15 - \log_5 3 = \log_5 \dfrac{15}{3} = \log_5 5 \Rightarrow 5^x = 5 \Rightarrow x = 1$

$\log_4 \dfrac{1}{4} = x \Rightarrow 4^x = \dfrac{1}{4} \Rightarrow x = -1$

$\log 100^{1/4} = \dfrac{1}{4}\log 100 = \dfrac{1}{4}(2) = \dfrac{1}{2}$

15. **(C)** The Cartesian Product of $T \times S$ consists of all ordered pairs, where the first element is chosen from T and the second element is chosen from S. The correct answer choice (6, 8) is not a member of $T \times S$. It is a member of $S \times T$.

16. **(C)** $P = \dfrac{A}{\left(1+\dfrac{r}{n}\right)^{nt}} = \dfrac{75{,}000}{\left(1+\dfrac{0.05}{2}\right)^{2(75)}} = 1847.06.$

An investment of \$1,847 will be worth \$75,000 in 75 years.

17. **(D)**

Coin 1	Coin 2	Coin 3	Coin 4	Coin 1	Coin 2	Coin 3	Coin 4
H	H	H	H	T	H	H	H
H	H	H	T	⇒T	H	H	T
H	H	T	H	⇒T	H	T	H
⇒H	H	T	T	T	H	T	T
H	T	H	H	⇒T	T	H	H
⇒H	T	H	T	T	T	H	T
⇒H	T	T	H	T	T	T	H
H	T	T	T	T	T	T	T

Of the 16 possibilities, 6 have exactly 2 heads. The probability is $\frac{6}{16}=\frac{3}{8}$

so the odds are 3:5.

Or: for each toss, there are two possibilities, heads or tails. With four tosses, there are $2^4 = 16$ possible outcomes.

The number of ways to choose two heads from 4 tosses is ${}_4C_2 = \frac{4!}{2!2!} = \frac{4\cdot3\cdot2\cdot1}{2\cdot1\cdot2\cdot1} = 6.$

18. The correct answer is 2. The irrational numbers are π and $\sqrt{5}$. Irrational numbers cannot be written as a quotient of two integers. The other four numbers can be written as a quotient of integers.

$$\sqrt{\frac{4}{25}} = \frac{2}{5},\ -.212 = -\frac{212}{1000},\ 5\frac{2}{7} = \frac{37}{7},\ \text{and } .1\overline{8} = \frac{17}{90}.$$

19. **(A)** The total number of points on Sheila's five exams is (5)(78) = 390. In order to attain a mean score of 84 on all the exams, she will need a total of (8)(84) = 672 points. Thus, she needs a total of 672 − 390 = 282 points on the next three exams. Finally 282 ÷ 3 = 94.

20. **(D)** Using the Pythagorean Theorem, $AB^2 + BC^2 = AC^2$. By substitution, $10^2 + BC^2 = 26^2$. Then $BC^2 = 676 - 100 = 576$, so $BC = 24$. The area of the triangle is given by the formula $\left(\frac{1}{2}\right)(AB)(BC) = \left(\frac{1}{2}\right)(10)(24) = 120.$

21. **(B)** To determine whether or not a graph represents a function, it is possible to apply the "vertical line test." If any vertical line to the graph intersects it in more than one point, the graph is not a function. The only graph for which a vertical line passes through more than one point is (B). A relation is a function if for any *x* there is one and only one *y*.

22. **(B)** $\% \text{ Markup} = \dfrac{\text{markup}}{\text{Cost}} \Rightarrow 0.8 = \dfrac{\text{markup}}{15} \Rightarrow \text{markup} = \$12$

Markup = selling price − cost ⇒ 12 = selling price − 15 ⇒ selling price = 27

$\% \text{ Markdown} = \dfrac{\text{markdown}}{\text{selling price}} \Rightarrow 0.3 = \dfrac{\text{markdown}}{27} \Rightarrow \text{markdown} = \$8.10$

Markdown = original price − new price ⇒ 8.10 = 27 − new price ⇒ new price = \$18.90

Markup = new selling price − cost ⇒ 18.90 − 15 = 3.90

$\% \text{ Markup} = \dfrac{\text{markup}}{\text{cost}} \Rightarrow \dfrac{3.90}{15} = 26\%$

23. **(B)** $-|-8-3| = -|-11| = -11$. The values of answer choices (A), (C), and (D) are 5, 0, and −5, respectively.

24. **(B)** In this example, $A' = \{\text{cat, goat, tiger}\}$.
Then, $A' \cap B = \{\text{cat, goat, tiger}\} \cap \{\text{dog, goat, pig, tiger}\} = \{\text{goat, tiger}\}$.

25. **(C)** Each leg is the radius r. $r^2 + r^2 = 16 \Rightarrow r^2 = 8$.

Area of circle is $\pi r2 = 8\pi$. Area of triangle $= \dfrac{1}{2}\left(\sqrt{8}\right)^2 = 4$.

Area of shaded region is $8\pi - 4$.

26. **(C)** The events are not mutually exclusive. They can occur at the same time. To be independent, Prob(Popcorn) · Prob(Soft drink) = Prob(Both)

Since $\left(\dfrac{40}{100}\right)\left(\dfrac{30}{100}\right) = \dfrac{12}{100}$ as $(0.4)(0.3) = 0.12$, having popcorn and a soft drink are independent.

Alternately: To be independent, Prob(Popcorn|Soft drink)
$= \text{Prob(Popcorn)} \Rightarrow \dfrac{12}{30} = \dfrac{40}{100}$ or $0.4 = 0.4$.

27. **(A)** The set $A \cup B$ contains all elements which belong to either set A or set B. Since all elements, which belong to set A or set B, also belong to C, the set $A \cup B$ is a subset of C.

28. **(D)** Given a conditional statement in the form "If P then Q," an equivalent statement is in the form "Not P or Q." In this example, P is the statement "Joan sings" and Q is the statement "I will play my guitar."

29. **(C)** $\% \text{ decrease} = \dfrac{\text{1975 Personnel} - \text{1980 Personnel}}{\text{1975 Personnel}}$

$$\text{Army: \% decrease} = \frac{784-777}{784} = 0.89\%$$

$$\text{Navy: \% decrease} = \frac{535-527}{535} = 1.50\%$$

$$\text{Air Force: \% decrease} = \frac{612-578}{612} = 5.56\%$$

$$\text{Marines: \% decrease} = \frac{196-188}{196} = 4.08\%$$

30. The answer is 800. The floor that measures 10 ft. by 20 ft. has an area of $10 \times 20 = 200$ sq. ft. The tiles with 36 sq. in. of area must measure 6 in. by 6 in. or $\frac{1}{2}$ ft. by $\frac{1}{2}$ ft. for $\frac{1}{4}$ sq. ft. of area. Because it would take four tiles to cover 1 sq. ft., $4 \times (200 \text{ sq. ft.}) = 800$ tiles would be needed to cover the entire floor.

31. **(C)** Prob(Washer or Dryer) = Prob(Washer + Prob(dryer) − Prob(Both)

$$\text{Prob}(\text{Washer or Dryer}) = \frac{55}{250} + \frac{45}{250} - \frac{25}{250} = \frac{75}{250} = \frac{3}{10} = 30\%$$

32. **(B)** Choosing two points that the line passes through: (200, 1,600), (300, 2,500), the slope of the line is $\dfrac{2{,}500-1{,}600}{300-200} = \dfrac{900}{100} = 9.$

33. **(D)** The domain refers to all allowable values of x. In this example, the only value(s) of x that are not allowed are those for which the denominator is zero. If $x + 3 = 0$, then $x = -3$. Thus, -3 is the only value that is not allowed in the domain.

34. **(A)** Since $x + c = a + b + c = 180°, x = a + b$
Since $y + b = a + b + c = 180°, y = a + c$
Since $z + a = a + b + c = 180°, z = b + c$
So $x + y - z = a + b + a + c - (b + c) = 2a + b + c - b - c = 2a$

35. **(C)** To find the median, first arrange the data in ascending order. The data will then appear as follows: 2, 7, 8, 10, 15, 24. The median will be the average of the two middle numbers. Thus, the median is $\dfrac{(8 + 10)}{2} = 9.$

36. **(A)** An argument is valid if given that the premises are true, then the conclusion *must* be true. Only answer choice (A) would satisfy this definition. Answer choice (B) is wrong because other objects besides trees may have brown leaves. Answer choice (C) is wrong because animals other than wolves may be vicious. Answer choice (D) is wrong because people may own only stocks, only bonds, both stocks and bonds, or neither stocks nor bonds.

37. **(C)** 10% of \$8 is \$0.80. So with the new prices, a customer averages \$7.20

$$x\% \text{ of } 7.20 = 10 \Rightarrow \frac{x}{100} \cdot 7.20 = 10 \Rightarrow 7.20x = 1000 \Rightarrow x = 138.89$$

The average customer must increase his spending by about 39%.

38. **(D)** The correct answer (D) has exactly two modes, namely 3 and 4. Answer choice (A) has a single mode of 5. Answer choice (B) has a single mode of 2. Answer choice (C) has three modes, namely 3, 4, and 5.

39. **(B)** To find the solution set of $f(x) > 2x$, we proceed as follows:

$$\frac{5x - 8}{2} > 2x$$

$$5x - 8 > 4x$$

$$-8 > -x$$

which implies $x > 8$.

40. **(A)** $(X \cup Y)'$ means the regions that are *not* included by X, by Y, or by both X and Y. Note that answer choice (C) is wrong because it includes region V, which is in all three of X, Y, and Z.

41. The correct answer is 2.92. The value of $\overline{X}$ is $(4 + 5 + 6 + 9 + 11) \div 5 = 7$.

The value of $\sum_{i=1}^{n} (X_i - \overline{X})^2$ can be found by computing

$$(4 - 7)^2 + (5 - 7)^2 + (6 - 7)^2 + (9 - 7)^2 + (11 - 7)^2$$

$$= 9 + 4 + 1 + 4 + 16 = 34.$$

Then $s = \sqrt{\frac{34}{4}} = \sqrt{8.5} \approx 2.92$

42. **(B)** Tax = 923 + 0.15(37,450 – 9,225) = 923 + 4,234 = \$5,157.

$$\%\text{Tax} = \frac{5{,}157}{37{,}450} = 0.1377 = 13.8\%$$

43. **(C)** Since g is a linear function, the ratio of the change in $g(x)$ values to the change in x values between any two points on the graph is constant. For the two given points (4, 6) and (10, 21), the constant ratio is $(21 - 6) \div (10 - 4) = 2.5$. Let $g(8) = k$. Combining this point with (4, 6), we can state that $(k - 6) \div (8 - 4) = 2.5$. Simplifying, we get $\frac{k-6}{4} = 2.5$. Then, multiplying both sides of the equation by 2.5, $k - 6 = 10$, so $k = 16$.

44. **(A)** x = selling price $\Rightarrow 0.06x = 16{,}250 \Rightarrow x = \$270{,}833$
Money at settlement: 270,833 – 16,250 = \$254,583

45. **(A)** Can: $\frac{0.40}{12} = \$0.0333/\text{ounce}$

48-ounce: $\frac{0.99}{48} = \$0.02063/\text{ounce}$

2-liter: $\frac{1.39}{67.6} = \$0.02056/\text{ounce}$

46. The correct answer 0.0048.
$(6 \times 10^3)(8 \times 10^{-1}) = 48 \times 10^{-4} = 4.8 \times 10^{-3} = 0.0048$

47. **(D)** The circumference of the wheel is:

$$C = 2\pi(1 \text{ ft.})$$
$$C = 2\left(\frac{22}{7}\right) = \frac{44}{7} \text{ ft.}$$

To find the number of revolutions the wheel takes, calculate:

$$5{,}280 \div \frac{44}{7} = 5{,}280 \times \frac{7}{44}$$
$$= 120 \times 7 = 840 \text{ revolutions}$$

48. **(A)** The mean and the median are each 6. For answer choice (B), the median is 6 but the mean is $\frac{43}{7}$. For answer choice (C), the mean is 6 but the median is 5.5. For answer choice (D), the median is 6 but the mean is 6.5.

49. **(B)** Rewriting $x + 3y = 4$ as $y = -\frac{1}{3}x + \frac{4}{3}$, we can identify the slope as $-\frac{1}{3}$. A line that is perpendicular to the graph of this line must have a slope that is the negative reciprocal of $-\frac{1}{3}$, which is 3. When answer choice (B) is rewritten as $y = 3x - 7$, the slope can be identified as 3. Note also that $y = 3x - 7$ contains the point $(2, -1)$. The slopes for answer choices (A), (C), and (D) are $-\frac{1}{3}$, -3, and $\frac{1}{3}$, respectively.

50. **(C)** Without seeing any of the data, we know nothing about the number and standard deviation. But since the slope of the boy line of best fit is slightly greater than the slope of the girl line of best fit, we can conclude that boys grow faster than girls in this school.

51. The correct answer is 5.33. By the Pythagorean Theorem, the height of the parallelogram is 4 and the area of the parallelogram is $4(8) = 32$. The area of the trapezoid is $\frac{1}{2}h(8+4) = 6h$. So $6h = 32$ and $h = \frac{16}{3} = 5.33$.

52. The correct answer is 22. $f(-7) = (4)(-7) - 1 = -29$ and $f(10) = (5)(10) + 1 = 51$. Then $-29 + 51 = 22$.

53. **(D)** There are 9 selections for the first spot on the shelf, 8 selections for the second spot, 7 selections for the third spot, and 6 selections for the fourth spot. The number of different ways is $(9)(8)(7)(6) = 3024$. This is a permutation of 9 items taken 4 at a time.

54. **(B)** The first \$2,000 is in the bank for 1 year: $A = 2000\left(1+\frac{0.03}{360}\right)^{360\cdot 1} = \$2,060.91$

The second \$3,000 is in the bank for half a year:

$$A = 3000\left(1+\frac{0.03}{360}\right)^{360\cdot 0.5} = \$3,045.34$$

Total = 2060.91 + 3045.34 = \$5, 106.25

55. **(A)** $f(x)$ is shifted 3 units to the left and 3 units down so $y = f(x + 3) - 3$.

56. **(B)**

	60-inch TV	32-inch TV	iPod	Nothing
X	2,000	500	230	0
Prob (X)	$\frac{1}{1,000}$	$\frac{2}{1,000}$	$\frac{5}{1,000}$	$\frac{992}{1,000}$

$$\text{Expected Value(win)} = 2000\left(\frac{1}{1000}\right) + 500\left(\frac{2}{1000}\right) + 230\left(\frac{5}{1000}\right) + 0\left(\frac{992}{500}\right)$$

$$= \frac{4150}{1000} = \$4.15$$

Person wins $\$4.15 - \$10.00 = -\$5.85$. A person can expect to lose \$5.85.

57. The correct answer is 5. The perimeter of the square is (4)(9) = 36, which is the same as the perimeter of the rectangle. The perimeter of a rectangle is twice the length plus twice the width. Twice the length is 26, so twice the width must be 10. Thus, the width is 5.

58. **(B)** I. True. The line is above the exponential curve.

    II. False. Ed's stock is worth \$4,000 at $t = 0$. Noreen's stock is worth 250(2) = \$500. The difference is \$3,500.

    III. True, Ed's stock is worth \$5,000 at $t = 4$. Noreen's stock is worth 250(32) = \$8,000.

60% of \$5,000 = \$3,000 and \$5,000 + \$3,000 = \$8,000.

59. The correct answer is $\frac{8}{9}$. Let $N = 0.\overline{8}$. Multiply both sides of the equation by 10 to get $10N = 8.\overline{8}$. Subtract $N = 0.\overline{8}$ from $10N = 8.\overline{8}$ to get $9N = 8$. Then $N = \frac{8}{9}$.

60. **(B)** Each handshake involves two people, so the number of handshakes for 20 people is given by the expression ${}_{20}C_2 = (20)(19) \div 2 = 190$.

PRACTICE TEST 2

CLEP College Mathematics

Also available at the REA Study Center (*www.rea.com/studycenter*)

This practice test is also available online at the REA Study Center. The CLEP College Mathematics test is only offered as a computer-based exam; therefore, we recommend that you take the online version of the practice test to receive these added benefits:

- **Timed testing conditions** – Gauge how much time you can spend on each question.
- **Automatic scoring** – Find out how you did on the test, instantly.
- **On-screen detailed explanations of answers** – Learn not just the correct answer, but also why the other answer choices are incorrect.
- **Diagnostic score reports** – Pinpoint where you're strongest and where you need to focus your study.

PRACTICE TEST 2

CLEP College Mathematics

(Answer sheets appear in the back of the book.)

TIME: 90 Minutes
60 Questions

Directions: An online scientific calculator will be available for the questions in this test.

Some questions will require you to select from among four choices. For these questions, select the BEST of the choices given.

Some questions will require you to type a numerical answer in the box provided.

Notes: (1) Unless otherwise specified, the domain of any function f is assumed to be the set of all real numbers x for which $f(x)$ is a real number.

(2) i will be used to denote $\sqrt{-1}$

(3) Figures that accompany questions are intended to provide information useful in answering the questions. All figures lie in a plane unless otherwise indicated. The figures are drawn as accurately as possible EXCEPT when it is stated in a specific question that the figure is not drawn to scale.

1. What is the domain of the function defined by $f(x) = \sqrt{-x+1} + 5$?

   (A) $\{x|x \geq 0\}$
   (B) $\{x|x \leq 1\}$
   (C) $\{x|0\ x \leq x \leq 1\}$
   (D) $\{x|x \geq -1\}$

2. A piece of merchandise is marked up and then, a week later, is marked down. Which of the following would give the most inexpensive price for the merchandise?

   (A) markup of 70%, then markdown of 70%
   (B) markup of 40%, then markdown of 40%
   (C) markup of 25%, then markdown of 25%
   (D) depends on the cost of the merchandise

3. A counting number with exactly two different factors is called a prime number. Which of the following pairs of numbers are consecutive prime numbers?

   (A) 27 and 29
   (B) 31 and 33
   (C) 41 and 43
   (D) 37 and 39

4. Which one of the following is equivalent to the statement "If roses are red, then the sky is blue"?

   (A) Roses are red and the sky is blue.
   (B) Roses are not red and the sky is blue.
   (C) Roses are red or the sky is blue.
   (D) Roses are not red or the sky is blue.

5. In the figure below, PQRS is a parallelogram. Find the value of y.

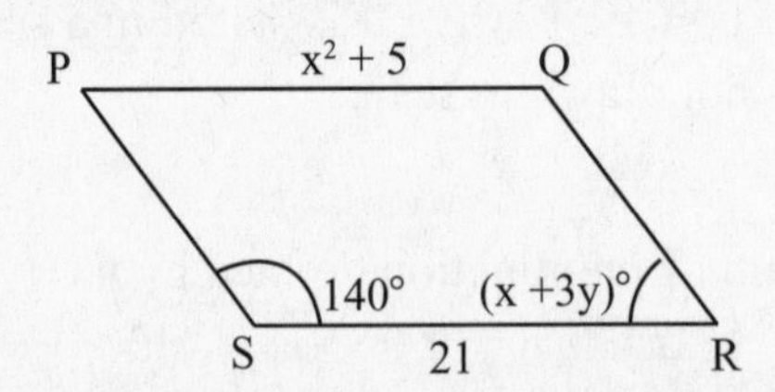

   (A) 4
   (B) 5
   (C) 12
   (D) 15

6. What is the percentage increase on a \$2.00 loaf of bread in 5 years if the rate of inflation is 2.48%?

(A) 11.6%
(B) 12.4%
(C) 13%
(D) 26%

7. If a and b are odd integers, which of the following must be an even integer?

I. $\frac{a+b}{2}$

II. $ab - 1$

III. $\frac{ab+1}{2}$

(A) I only
(B) II only
(C) I and II only
(D) II and III only

8. Arrange the following data sets by standard deviation, from smallest to largest.

I. {100, 102, 104, 116, 118, 120}
II. {1000, 1002, 1004, 1006, 1008, 1010}
III. {20, 20, 20, 30, 30, 30}

(A) III, II, I
(B) I, III, II
(C) I, II, III
(D) II, I, III

9. What is the contrapositive of the statement "If today is Tuesday, then we are going to Chicago"?

(A) If today is not Tuesday, then we are not going to Chicago.
(B) If we are going to Chicago, then today is Tuesday.
(C) If today is Tuesday, then we are not going to Chicago.
(D) If we are not going to Chicago, then today is not Tuesday.

10. If $\log_{10} x = 3$ and $\log_{10} .01 = y$, what is the value of $x + y$?

11. A new car loan is advertised as 5.2% APR compounded continuously. What is the effective rate?

(A) 5.2%
(B) 5.31%
(C) 5.34%
(D) 5.55%

12. Five hundred people were surveyed concerning their favorite snack. The pie chart shown below is a summary of the data collected.

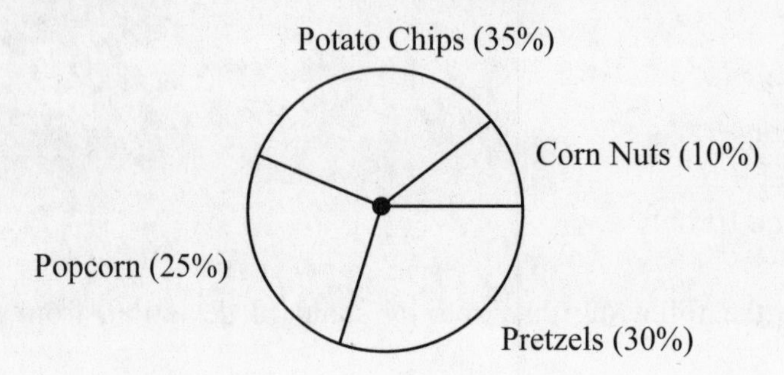

What is the combined total of people who chose either pretzels or potato chips as their favorite snack?

(A) 425
(B) 325
(C) 300
(D) 250

13. The area of a trapezoid is given by the formula $A = \left(\frac{1}{2}\right)(h)(b_1 + b_2)$, where h represents the height, b_1 represents one base, and b_2 represents the other base. What is the area of a trapezoid in which the height is 5 and the mean value of the two bases is 7?

14. Which one of the graphs below represents the inverse of the function $y = 3x + 4$?

(A)

(C)

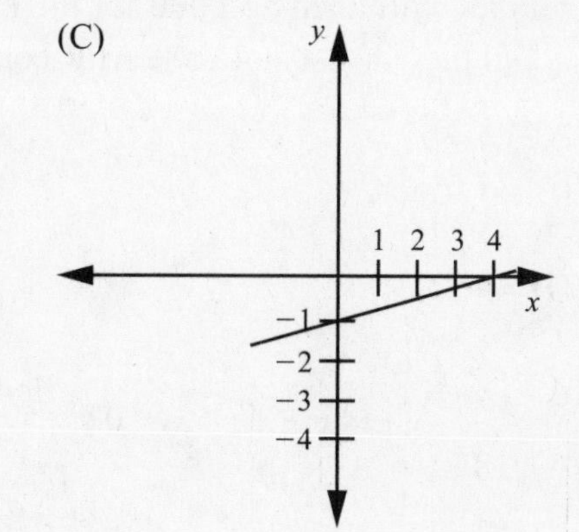

(B)

(D)

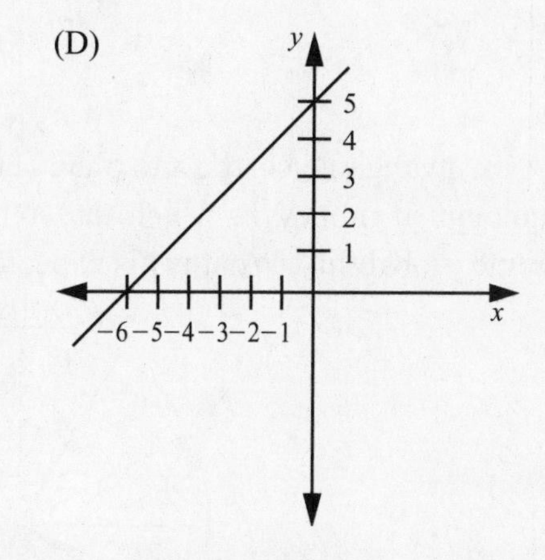

15. Given sets T, V such that $T \cap V = \{3, 7\}$ and $T \cup V = \{1, 3, 5, 7, 9, 11\}$, which one of the following could *NOT* represent T?

(A) $\{1, 3\}$
(B) $\{3, 7\}$
(C) $\{1, 3, 7\}$
(D) $\{3, 5, 7\}$

16. At the time of their 10th anniversary, the Triplers invest a sum of money that will be used on their 25th anniversary for a world cruise. They estimate that they will need \$75,000 for the cruise. What percentage of the \$75,000 should they invest at 4.65% APR compounded daily?

    (A) 23.3%
    (B) 27.6%
    (C) 31.3%
    (D) 49.8%

17. If $f(x)=\begin{cases} x^2-x-2, & x\geq 0 \\ -|4x-2|, & x<0 \end{cases}$, find $f(4) - f(-4)$

    (A) 28
    (B) 24
    (C) −4
    (D) −8

18. A carnival game costs \$8 to play. The player spins the spinner and wins the amount of money in which the spinner lands. Each landing spot has the same probability. What is his expected value for the game?

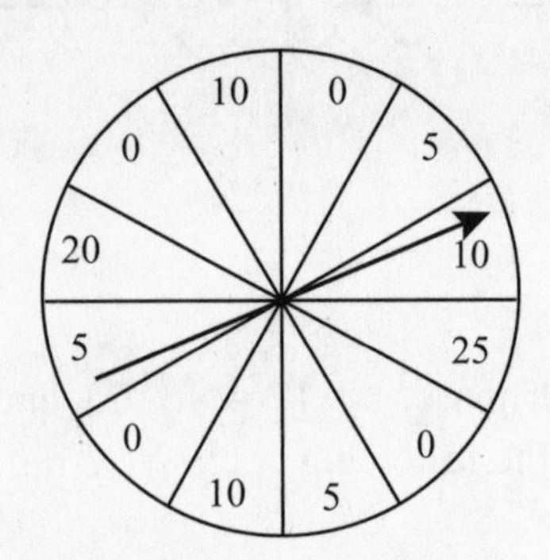

    (A) win \$7.50
    (B) lose \$0.50
    (C) break even
    (D) lose \$2.00

19. If $f(x) = 2x^2 + 5$ and $g(x) = -x - 6$, what is the value of $f(g(-1))$?

    (A) −93
    (B) −45
    (C) 13
    (D) 55

20. Using only the digits 1, 2, 3, 4, 5, how many different four-digit numbers are possible if the left-most digit is odd and the right-most digit is even? Repetition of digits is allowed.

(A) 150
(B) 100
(C) 48
(D) 36

21. Randy purchased a tract of land for $80,000. He later sold it at $115,000. If he has to pay a 15% Capital Gains Tax, what is his profit on the sale on the land?

(A) $17,750
(B) $29,750
(C) $40,250
(D) $52,250

22. What is the area of this triangle?

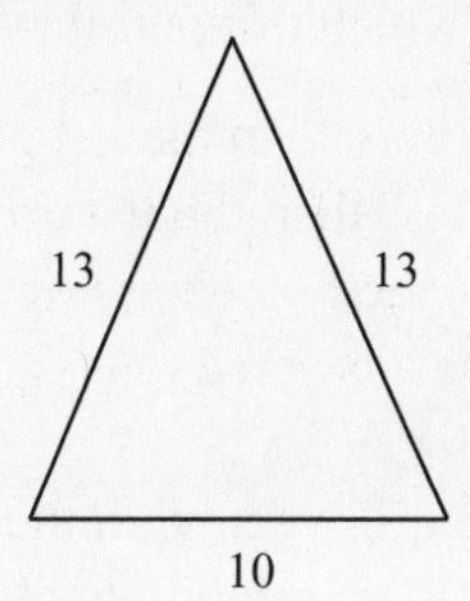

(A) 60
(B) 65
(C) 120
(D) 130

23. Let P, Q, and R each represent false statements. Which one of the following statements is also false?

(A) P implies (Q and R)
(B) (P and Q) or (not R)
(C) P or (Q implies not R)
(D) P and (not Q or R)

24. In the British monetary system, \$1 = 0.67 pounds. 1 liter is equivalent to 0.26 gallons. If the price of gas is \$3 per gallon, how much would it be in pounds per liter?

(A) 0.52
(B) 0.86
(C) 1.16
(D) 7.73

25. Which of the following have the same values?

I. $\log_2 8$ II. $\dfrac{1}{\log_8 2}$ III. $\log_{\frac{1}{2}} \dfrac{1}{8}$

(A) I and II only
(B) I and III only
(C) all do
(D) none do

26. A new car dealer offers a 6% rebate on the list price of a new car. If the tax on the selling price is 6%, which of the following is true?

(A) The buyer will spend more than the list price.
(B) The buyer will spend less than the list price.
(C) The buyer will spend exactly the list price
(D) The question cannot be answered unless the list price is known.

27. In a school district's budget, 0.2 cents on the dollar goes to rebinding old textbooks. If \$20,000 went to rebinding old books, what is the school district's budget?

(A) \$1 million
(B) \$10 million
(C) \$100 million
(D) \$1 billion

28. Let P and Q represent statements. In which one of the following is P a necessary condition for Q?

(A) P implies Q
(B) P and Q
(C) Q implies P
(D) Q or P

29. A deck of cards contains 4 suits (clubs, diamonds, hearts, and spades). Each suit has 13 cards: ace, 2, 3, 4, 5, 6, 7, 8, 9, 10, and the face cards, (Jack, Queen, King). What is the probability of choosing a spade or a face card?

(A) $\frac{11}{26}$

(B) $\frac{25}{52}$

(C) $\frac{3}{13}$

(D) $\frac{1}{2}$

30. Look at the following Venn Diagram, where J and K represent sets.

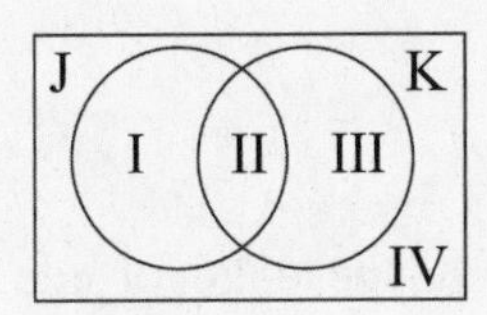

Which region(s) represent K′?

(A) I and IV
(B) Only IV
(C) Only I
(D) III and IV

31. If the graph of a function $f(x)$ crosses the x-axis twice, which one of the following statements is true?

(A) There are exactly two distinct real values for $f(x) = 0$.
(B) There is exactly one real value for $f(x) = 0$.
(C) There are at most two distinct real values for $f(x) = 0$.
(D) There are at least two distinct real values for $f(x) = 0$.

32. Ryan is a pilot and has an annual income of \$120,000. He is laid off and finds a new job that pays him 30% less. He pays federal tax according to the chart below. When he starts the new job, what is the percentage decrease in the amount of tax he pays?

If taxable income is between	The tax due is
0 – \$9,225	10% of the taxable income
\$9,226 – \$37,450	\$923 + 15% of the amount over \$9,225
\$37,451 – \$90,750	\$5,156 + 25% of the amount over \$37,450
\$90,751 – \$189,300	\$18,481 + 28% of the amount over \$90,750

(A) 3%
(B) 37.03%
(C) 59.5%
(D) 81.5%

33. Triangles ABC and $A'B'C'$ are similar right triangles. Find the length of side AC.

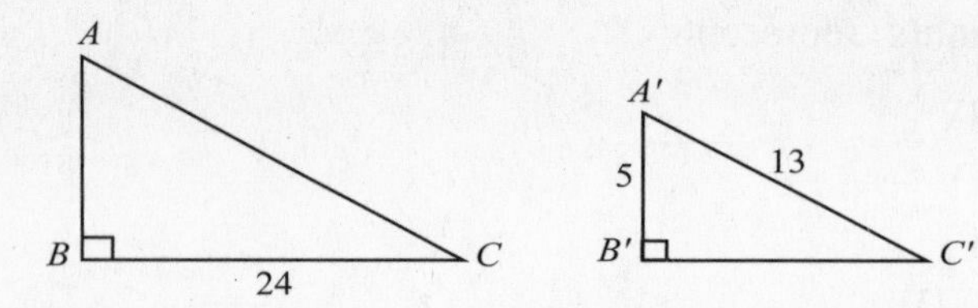

(A) 10
(B) 12
(C) 13
(D) 26

34. Professor Leonardo gave a quiz to the 20 students in her class. After calculating the mean grade to be 80, she realized that she mistakenly wrote 95 for one of her students. If this student's test score was really 55, what was the actual mean grade for the class?

(A) 78
(B) 77
(C) 76
(D) 75

35. There are 4 men and 6 women in a room. Two people will be randomly selected. What is the probability that both will be men?

(A) $\frac{2}{3}$

(B) $\frac{2}{5}$

(C) $\frac{2}{11}$

(D) $\frac{2}{15}$

36. Let $S = \{x \mid x$ is an integer greater than $2\}$, let $T = \{x \mid x$ is an odd negative integer$\}$, and let $V = \{x \mid x$ is a non-zero integer with an absolute value less than $4\}$. Which one of the following is equivalent to the empty set?

(A) $T \cap V$
(B) $S \cap V$
(C) $S \cap T$
(D) $T \cup V$

37. Which one of the following is an irrational number?

(A) $\sqrt{.0025}$

(B) $.\overline{124}$

(C) .26226222622226....

(D) $\frac{11}{7}$

38. For a group of data, arranged in ascending order, the median is located midway between the twelfth and thirteenth numbers. How many data are in this group?

39. Which one of the following functions has a range of all numbers except zero?

(A) $f(x) = x - 1$

(B) $f(x) = \frac{1}{x}$

(C) $f(x) = \frac{1}{\sqrt{x}}$

(D) $f(x) = -|x|$

40. Mark invests $40,000 in a bank at 3.5% APR compounded quarterly. Three years later, he withdraws half of his money for a new car, and keeps the remainder in the bank at the same interest rate. Three years later, he will withdraw all of his money for another new car. How much money will he withdraw?

(A) $24,420
(B) $24,585
(C) $24,651
(D) $26,336

41. Let $F = \{3, 6, 9\}$, $G = \{4, 6, 8\}$, and $H = \{6, 8, 9\}$. The ordered pair (3, 6) belongs to which of the following Cartesian products?

(A) $F \times G$ and $F \times H$
(B) Only $F \times G$
(C) Only $F \times H$
(D) $F \times H$ and $G \times H$

42. If m and n are consecutive integers, and $m < n$, which one of the following statements is always true?

(A) $n - m$ is even.
(B) m must be odd.
(C) $m^2 + n^2$ is even.
(D) $n^2 - m^2$ is odd.

43. The points $(-2, 6)$ and $(1, 11)$ lie on the graph of a linear function. Which one of the following points must also lie on the graph of this function?

(A) $(2, 14)$
(B) $(3, 15)$
(C) $(4, 16)$
(D) $(5, 17)$

44. Which one of the following is the negation for the statement "Some women enjoy shopping"?

(A) Some women do not enjoy shopping.
(B) At least one woman enjoys shopping.
(C) No women enjoy shopping.
(D) All women enjoy shopping.

45. The graph of two linear equations in x and y contains two perpendicular lines. Which system below could represent the equations of these two lines?

(A) $4x + 7y = 11$
$8x + 14y = 13$

(B) $5x + 11y = 17$
$11x + 5y = 20$

(C) $3x + 7y = 10$
$14x - 6y = 15$

(D) $9x + 2y = 19$
$18x - 4y = 21$

46. What is the smallest positive number such that when it is divided by 9 or by 12, the remainder is 1?

47. The Castors invest \$1,000 in a Christmas account on September 1. The account gains interest at 4% APR compounded monthly. On October 1, they add another \$1,000 to the account and on November 1, they add another \$1,000 to the account On December 1, they withdraw the money to shop for presents. To the nearest dollar, how much interest did the account gain?

(A) \$10
(B) \$20
(C) \$246
(D) \$251

48. Let J represent any non-empty set. Consider the following group of combinations of sets:

$$J \cup \emptyset,\ J \cap \emptyset,\ J - J',\ J \cap J',\ J - \emptyset,\ \emptyset - J$$

How many of these six combinations are equivalent to Ø?

49. For which one of the following data groups is the mean and median identical, but there is no mode?

(A) 2, 2, 4, 4, 4, 6, 6
(B) 2, 4, 6, 8, 10, 12
(C) 2, 5, 7, 9, 10, 13
(D) 2, 3, 5, 5, 6, 7, 7

50. A line and an exponential function whose equation is given by $y = 2^{x} + 1$ both pass through the points (0, 4) and (4, 32). What is the difference between the two functions when evaluated at $x = -3x$?

(A) 13
(B) 21
(C) $\frac{67}{4}$
(D) $\frac{69}{4}$

51. On a bookshelf, there are 25 books, 10 of which are red, and the remaining books are green. Twelve of these books are math books of which 4 are red. How many green non-math books are there?

(A) 7
(B) 8
(C) 11
(D) 15

52. Four less than three times x is greater than 6. Find all values of x.

(A) $x < \frac{10}{3}$
(B) $x > \frac{10}{3}$
(C) $x < 5$
(D) $x > \frac{2}{3}$

53. A fruit distributor knows that during the month of October, the weight of apples is normally distributed with a mean of 0.64 lb. and a standard deviation of 0.14 lb. In a shipment of 5,000 apples, how many can be expected to weigh more than a half a pound?

(A) 1,700
(B) 3,300
(C) 4,200
(D) 4,750

54. Doug bought shares of stock for $9,000. The value of the stock increased by 6% a year over 5 years. At that time, Doug sold the stock and had to pay a 15% Capital Gains Tax on the profit from selling the stock. After paying the tax, what was Doug's profit on the sale of the stock to the nearest dollar?

(A) $945
(B) $1,237
(C) $2,295
(D) $2,587

55. Calculate how much interest a borrower must pay on a 33-month loan of $20,250 if the interest rate is 6.5% APR compounded continuously.

(A) $3,453
(B) $3,573
(C) $3,829
(D) $3,963

56. The formula for the area of a circle is $A = \pi r^2$, where r represents the radius. How is the area of a given circle affected if the radius is doubled?

(A) The area is doubled.
(B) The area is tripled.
(C) The area is quadrupled.
(D) The change in area cannot be determined without knowing the value of the radius.

57. Suppose the probability that it will rain today is .8 and the probability that Ken will go bowling today is .6. Assuming that these events are independent, what is the probability that Ken will *NOT* go bowling today and it will also *NOT* rain today?

(A) .70
(B) .40
(C) .20
(D) .08

58. Which one of the following functions shows that both the domain and the range are all non-zero numbers?

(A) $f(x) = \sqrt{x}$

(B) $f(x) = \frac{2}{x}$

(C) $f(x) = |x|$

(D) $f(x) = x - 3$

59. An equilateral triangle and a rectangle have identical perimeters. Each side of the triangle is 13 and the width of the rectangle is 8. What is the rectangle's length?

60. A scatterplot is shown below along with the line of best fit. If the point (1, 6) is added, what would happen to the strength of the association and the slope of the line of best fit?

(A) stronger association and steeper slope
(B) stronger association and more shallow slope
(C) weaker association and steeper slope
(D) weaker association and more shallow slope

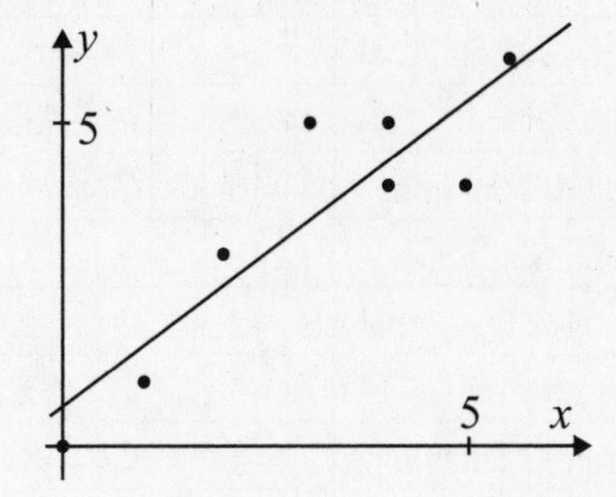

PRACTICE TEST 2

Answer Key

#	Answer	Chapter	#	Answer	Chapter
1	B	3. Algebra & Functions	31	A	3. Algebra & Functions
2	A	6. Financial Math	32	B	6. Financial Math
3	C	2. Numbers	33	D	7. Geometry
4	D	8. Logic & Sets	34	A	5. Statistics & Data Analysis
5	C	7. Geometry	35	D	4. Probability & Counting
6	C	6. Financial Math	36	C	8. Logic & Sets
7	B	2. Numbers	37	C	2. Numbers
8	D	5. Statistics & Data Analysis	38	24	5. Statistics & Data Analysis
9	D	8. Logic & Sets	39	B	3. Algebra & Functions
10	998	3. Algebra & Functions	40	C	6. Financial Math
11	C	6. Financial Math	41	A	8. Logic & Sets
12	B	5. Statistics & Data Analysis	42	D	2. Numbers
13	35	7. Geometry	43	C	3. Algebra & Functions
14	C	3. Algebra & Functions	44	C	8. Logic & Sets
15	A	8. Logic & Sets	45	C	3. Algebra & Functions
16	D	6. Financial Math	46	37	2. Numbers
17	A	3. Algebra & Functions	47	B	6. Financial Math
18	B	4. Probability & Counting	48	3	8. Logic & Sets
19	D	3. Algebra & Functions	49	B	5. Statistics & Data Analysis
20	A	4. Probability & Counting	50	D	3. Algebra & Functions
21	B	6. Financial Math	51	A	4. Probability & Counting
22	A	7. Geometry	52	B	3. Algebra & Functions
23	D	8. Logic & Sets	53	C	5. Statistics & Data Analysis
24	A	2. Numbers	54	D	6. Financial Math
25	C	3. Algebra & Functions	55	D	6. Financial Math
26	B	6. Financial Math	56	C	7. Geometry
27	B	6. Financial Math	57	D	4. Probability & Counting
28	C	8. Logic & Sets	58	B	3. Algebra & Functions
29	A	4. Probability & Counting	59	11.5	7. Geometry
30	A	8. Logic and Sets	60	D	5. Statistics and Data Analysis

PRACTICE TEST 2

Detailed Explanations of Answers

1. **(B)** The only restriction for the domain is that $-x + 1$ must be greater than or equal to zero.

   $-x + 1 \geq 0$

   so $1 \geq x$, which means $x \leq 1$.

2. **(A)** Let c be the cost of the merchandise.
   Price of A: $(1.7c)(0.3) = 0.51c$
   Price of B: $(1.4c)(0.6) = 0.84c$
   Price of C: $(1.25c)(0.75) = 0.9375c$

3. **(C)** To test whether a number, N, is prime, we need to test if N is divisible by any of the prime numbers $\{2, 3, 5, 7, 11, 13, \ldots\}$ up to the largest natural number, k, whose square is less than or equal to the number we are testing, N. If N is divisible by any of the prime numbers $P \leq k$, where $k^2 \leq N$, then N is not a prime number. If N is not divisible by any of the prime numbers $P \leq k$, where $k^2 \leq N$, then N is a prime number. For example, to test whether 29 is a prime number, we need to test if 29 is divisible by any of the prime numbers starting with 2 and up to 5. That is, we test if 29 is divisible by 2, 3, or 5, since $6^2 = 36$ is > 29. Since 29 is not divisible by any of these primes, 29 is a prime number.

   Since all the prime numbers, except 2, are odd, it follows that the difference between any two consecutive prime numbers is 2. Thus, each of the pairs of numbers given in the answer choices as possible answers are two consecutive odd numbers.

   To answer this question, we need to test if any of the pairs of numbers given in the answer choices is a pair of prime numbers. Testing these pairs of numbers yields:

   (A) 27 and 29. Since $6^2 = 36$, and since $36 > 27$, and $36 > 29$, we need to test if 27 or 29 is divisible by any of the prime numbers less than or equal to 5. That is, we check if 27 or 29 is divisible by 2, 3, or 5. Since 27 is divisible by 3, then 27 and 29 is not a pair of consecutive prime numbers.

(B) 31 and 33. Again, $6^2 = 36$, $36 > 31$, and $36 > 33$. Hence, we need to test if 31 or 33 is divisible by any of the prime numbers less than or equal to 5. That is, we check if 31 or 33 is divisible by any of the primes 2, 3, or 5. Since 33 is divisible by 3, then 31 and 33 is not a pair of consecutive prime numbers.

(C) 41 and 43. Since $7^2 = 49$, $49 > 41$, and $49 > 43$, it follows that we need to test if 41 or 49 is divisible by any of the prime numbers less than or equal to 6. That is, we check if 41 or 43 is divisible by any of the primes 2, 3, or 5. Since 41 is not divisible by any of these three primes and 43 is not divisible by any of these three primes either, it follows that 41 and 43 is a pair of consecutive prime numbers.

(D) 37 and 39. Since 39 is divisible by 3, then 37 and 39 is not a pair of consecutive prime numbers.

4. **(D)** When a statement is in the form "If P, then Q," the equivalent statement is in the form "Not P or Q." Let P represent the statement "Roses are red" and let Q represent the statement "The sky is blue." Then "Not P" is represented by "Roses are not red."

5. **(C)** In a parallelogram, opposite sides are congruent so $x^2 + 5 = 21 \Rightarrow$ $x = 4$. Also consecutive angles are supplementary (adding to 180°) so $140 + 4 + 3y = 180 \Rightarrow 3y = 36 \Rightarrow y = 12$

6. **(C)** $P = 2$, $r = 0.0248$, $n = 1$ (since inflation is an annual rate), $t = 5$

$$A = P\left(1+\frac{r}{n}\right)^{nt} = 2\left(1+\frac{0.0248}{1}\right)^{1(5)} = 2.26$$

In 5 years, the price of bread will be \$2.26 (going up \$0.26)

$$\text{Percent increase} = \frac{0.26}{2.00} = 13\%$$

7. **(B)** We can express odd integers as $a = 2x + 1$ and $b = 2y + 1$ where x and y are integers.

I. $$\frac{a+b}{2} = \frac{(2x+1)+(2y+1)}{2}$$

$$= \frac{2x+2y+2}{2}$$

$= x + y + 1,$

which is not necessarily even.

II. $ab - 1 = (2x + 1)(2y + 1) - 1$
$= (4xy + 2x + 2y + 1) - 1$
$= 2(2xy + x + y),$

which is always even (divisible by 2).

III. $\dfrac{ab+1}{2} = \dfrac{(4xy + 2x + 2y + 1) + 1}{2}$
$= 2xy + x + y + 1,$

which is not necessarily even.

8. **(D)** It is not necessary to find the standard deviation mathematically to answer the question. II has a mean of 1005 and the data has the smallest deviation from the mean. I has a mean of 110 and the data is farther from the mean than in II. III has a mean of 25 and the data is farthest from the mean.

9. **(D)** Given a statement in the form "If P then Q," the contrapositive is the statement "If not Q, then not P." In this example, P represents "Today is Tuesday" and Q represents "we are going to Chicago."

10. The correct answer is 998. $\log_{10} x = 3$ means $x = 10^3 = 1000$ and $\log_{10} .01 = y$ means $10^y = .01 = 10^{-2}$, so $y = -2$. Then $x + y = 1000 - 2 = 998$.

11. **(C)** The principal amount makes no difference. Use \$100 to make the numbers easy.
$A = Pe^{rt} = 100(2.718)^{0.052(1)} = 105.34$
$I = A - P = 105.34 - 100 = 5.34$
So the compounded interest is \$5.34 which means the effective interest rate is 5.34%.

12. **(B)** The total percent of people who chose either pretzels or potato chips as their favorite snack is 30% + 35% = 65%. This percent corresponds to (.65)(500) = 325 people.

13. The correct answer is 35. The easiest way to calculate this answer is to rewrite the area formula as $A = \left(\frac{1}{2}\right)(b_1 + b_2)(h)$. The expression $\left(\frac{1}{2}\right)(b_1 + b_2)$ represents the mean of b_1 and b_2. Then the area is $(7)(5) = 35$.

14. **(C)** The inverse of the function $f(x) = 3x + 4$ is obtained by replacing y by x and x by y.

$y = 3x + 4$

becomes

$x = 3y + 4 \Rightarrow 3y = x - 4 \Rightarrow y = \dfrac{x-4}{3}$

The graph of $y = \dfrac{x-4}{3}$ is given below:

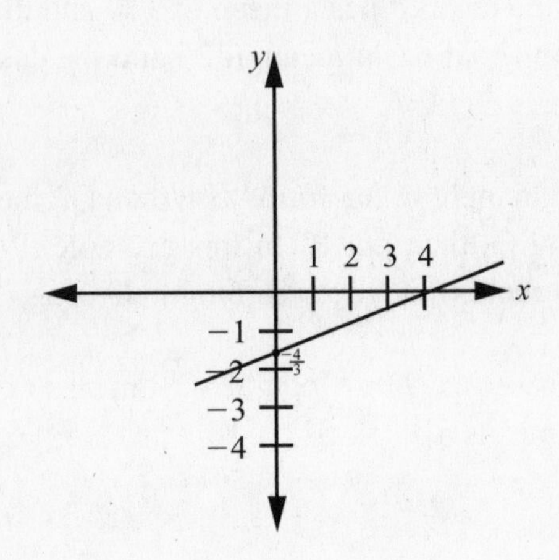

15. **(A)** Since $T \cap V = \{3, 7\}$, T *must* contain at least the elements 3 and 7. Answer choice (A) does not contain the element 7.

16. **(D)**

$$P = \frac{A}{\left(1+\frac{r}{n}\right)^{nt}} = \frac{75{,}000}{\left(1+\frac{0.0465}{360}\right)^{360(15)}} = \$37{,}339$$

They should invest \$37,339. This is $\dfrac{37{,}339}{75{,}000} = 49.8\%$.

17. **(A)** $f(4) = 16 - 4 - 2 = 10$
$f(-4) = -|-16 - 2| = -18$
$f(4) - f(-4) = 10 - (-18) = 28$

18. **(B)**

x	25	20	10	5	0
Prob(*x*)	$\frac{1}{12}$	$\frac{1}{12}$	$\frac{3}{12}$	$\frac{3}{12}$	$\frac{4}{12}$

Expected value:

$$25\left(\frac{1}{12}\right)+20\left(\frac{1}{12}\right)+10\left(\frac{3}{12}\right)+5\left(\frac{3}{12}\right)+0\left(\frac{4}{12}\right)=\frac{90}{12}=7.50$$

You win \$7.50 but it costs \$8 to play, so you lose 50 cents.

19. **(D)**

$g(-1) = -(-1) - 6 = 1 - 6 = -5.$

Then, $f(-5) = (2)(-5)^2 + 5 = (2)(25) + 5 = 55.$

20. **(A)** The number of choices for the left-most digit is 3, since there are three odd digits (1, 3, 5). The number of choices for the right-most digit is 2, since there are two even digits (2, 4). For each of the other two digits there are 5 choices, since repetition of any of the five digits is allowed. Then $(3)(5)(5)(2) = 150$ represents the number of possibilities when creating a four-digit number.

21. **(B)** Capital gains tax $= 0.15(115{,}000 - 80{,}000) = 0.15(35{,}000) = \$5{,}250$.
Profit $= 115{,}000 - 80{,}000 - 5{,}250 = \$29{,}750$.

22. **(A)** Draw the perpendicular bisector h as shown. The base of each small triangle is 5. By the Pythagorean Theorem, $5^2 + h^2 = 13^2$ and $h = 12$.

So the area of the triangle is $\frac{1}{2}(10)(12) = 60$.

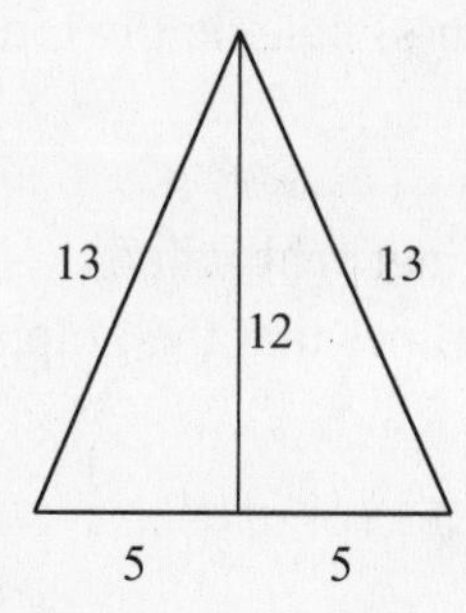

23. **(D)** Since P is false, a compound statement with false and any truth value must be false. The statement "Not Q or R" is actually true, but "false and true" is equivalent to false. In answer choice (A), it would read: "false implies false," which is true. In answer choice (B), it would read: "false or true," which is true. In answer choice (C), it would read: "false or (false implies true)," which becomes "false or true," which is true. Recall that "false implies any truth value" is always true.

24. **(A)** $\frac{\$3.00}{\text{gallon}} \quad \frac{0.26 \text{ gallon}}{1 \text{ liter}} \quad \frac{0.67 \text{ pounds}}{\$1.00} \quad \frac{0.52 \text{ pounds}}{\text{liter}}$

25. **(C)** I. $\log_2 8 = x \Rightarrow 2^x = 2^3 \Rightarrow x = 3$

II. $\log_8 2 = x \Rightarrow 8^x = 2 \Rightarrow 2^{3x} = 2^1 \Rightarrow 3x = 1 \Rightarrow x = \frac{1}{3}$ so $\frac{1}{\log_8 2} = \frac{1}{1/3} = 3$

III. $\log_{\frac{1}{2}} \frac{1}{8} = x \Rightarrow \left(\frac{1}{2}\right)^x = \left(\frac{1}{2}\right)^3 \Rightarrow x = 3$

26. **(B)** If the list price is x, the buyer will spend $x(0.94)(1.06) = 0.9964x$ This is just slightly less than the list price. If the car lists for \$40,000, the buyer would spend \$39,856.

27. **(B)** $\frac{0.2}{100}x = 20{,}000 \Rightarrow 0.2x = 2{,}000{,}000 \Rightarrow x = \frac{2{,}000{,}000}{0.2} =$ $10{,}000{,}000 =$ \$10 million.

28. **(C)** "P is a necessary condition for Q" is equivalent to "If Q, then P," which means "Q implies P."

29. **(A)** There are 52 cards altogether. There are 13 spades and 12 face cards. But 3 of the face cards are also spades so there are 22 cards that are either a spade or a face card. Answer $= \frac{22}{52} = \frac{11}{26}$.

30. **(A)** K′ means the region(s) outside the circle shown as K, which would be both I and IV.

31. **(A)** The number of distinct real values for $f(x) = 0$ can be found by looking at the number of times that the graph of $f(x)$ crosses the x-axis.

32. **(B)** Income $= \$120{,}000$: Tax $= \$18{,}481 + 0.28(120{,}000 - 90{,}750) = \$26{,}771$

$$30\% \text{ decrease income} = 120{,}000 - 0.3(120{,}000) = \$84{,}000$$

$$\text{Income} = \$84{,}000: \text{Tax} = \$5{,}156 + 0.25(84{,}000 - 37{,}450) = \$16{,}794$$

$$\% \text{ Decrease} = \frac{26{,}671 - 16{,}794}{26{,}671} = 37.3\%$$

33. **(D)** The sides of similar triangles are proportional. Using the Pythagorean Theorem, the length of $B'C'$ is:

$$5^2 + (B'C')^2 = 13^2$$
$$25 + (B'C')^2 = 169$$
$$(B'C')^2 = 144$$
$$B'C' = 12$$

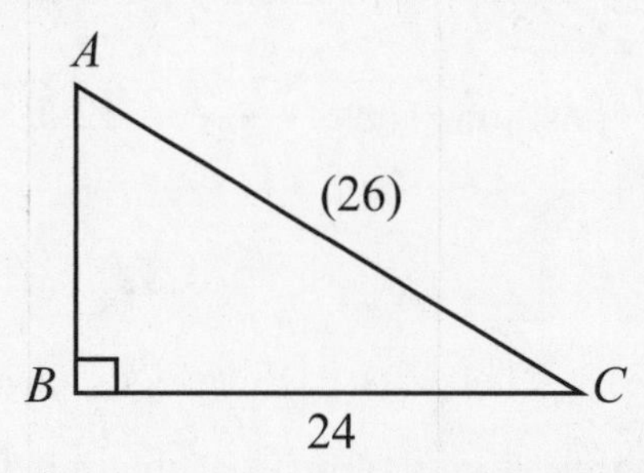

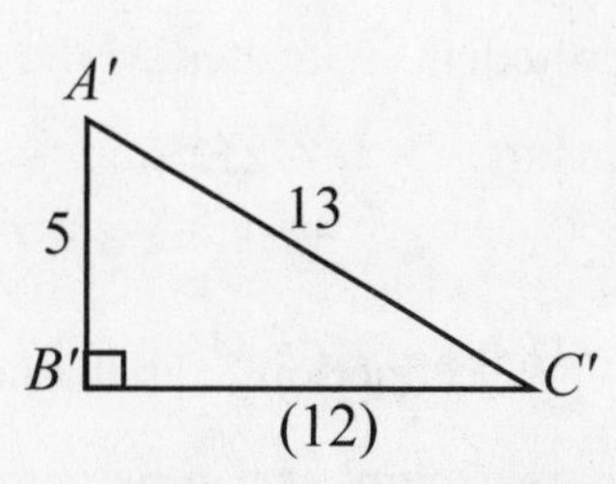

BC and $B'C'$ are corresponding sides of similar triangles in a ratio of 24:12 or 2:1. Similarly, side AC and $A'C'$ have the same ratio; AC must be twice as long as $A'C'$. In other words, $AC = 2(13) = 26$.

34. **(A)** The original total points for the class was $(80)(20) = 1600$. Since one of the students was erroneously given 40 extra points, the total points for the class should have been 1560. The correct mean for the class is $1560 \div 20 = 78$.

35. **(D)** The number of combinations of selecting two men is ${}_4C_2 = \frac{(4)(3)}{2!} = 6$. The number of combinations of selecting any two people is ${}_{10}C_2 = \frac{(10)(9)}{2!} = 45$. Then the probability is given by $\frac{6}{45} = \frac{2}{15}$.

36. **(C)** $S = \{3, 4, 5, 6, \ldots\}$, $T = \{-1, -3, -5, -7, \ldots\}$, and $V = \{-1, -2, -3, 1, 2, 3\}$. We note that S and T have no common elements, so their intersection is the empty set. $T \cap V = \{-1, -3\}$ and $S \cap V = \{3\}$.

37. **(C)** An irrational number is one that cannot be written in the form $\frac{P}{Q}$, where P and Q are integers. Any decimal that does not end or repeat is always an irrational number. Answer choice (C) shows a definite pattern, but the number of 2's between the 6's never repeats. Answer choice (A) can be written as .05 or $\frac{1}{20}$. Answer choice (B) can be written as $\frac{124}{999}$, and answer choice (D) is already written in the form $\frac{P}{Q}$, where $P = 11$ and $Q = 7$.

38. The correct answer is 24. For a group of n data, arranged in ascending order, the median is located in the $\left(\frac{n+1}{2}\right)^{th}$ position. In this example, the median is located in the 12.5^{th} position. Then $\frac{n+1}{2} = 12.5$, which leads to $n + 1 = 25$. So, $n = 24$.

39. **(B)** The fraction $\frac{1}{x}$ is defined except when $x = 0$. Since 1 divided by a non-zero number can be any non-zero value, the range of this function is all numbers except zero. For answer choice (A), the range is all numbers. For answer choice (C), the range is all positive numbers. For answer choice (D), the range is all non-positive numbers.

40. **(C)**

$$40{,}000\left(1+\frac{0.035}{4}\right)^{4(3)} = \$44{,}408 \text{ so he withdraws \$22,204, leaving \$22,204.}$$

$$\$22{,}204\left(1+\frac{0.035}{4}\right)^{4(3)} = \$24{,}651.$$

41. **(A)** The Cartesian product of any two sets is formed using ordered pairs in which the first element comes from the first set and the second element comes from the second set. Since the number 3 comes from set F and the number 6 comes from each of sets G and H, the ordered pair (3, 6) belongs to both $F \times G$ and $F \times H$.

42. **(D)** If m and n are consecutive integers, and $m < n$, it follows that $n = m + 1$

Now, we can check each of the answer choices (A) through (D) as follows:

(A) $n - m = (m + 1) - m = m + 1 - m = 1$,
which is odd. Thus, the statement in answer choice (A) is false.

(B) Since no specific information is given about the integer m, m can be an odd integer or an even integer. So, the statement in answer choice (B) is false.

(C) $$\begin{aligned} m^2 + n^2 = m^2 + (m + 1)^2 &= m^2 + m^2 + 2m + 1 \\ &= 2m^2 + 2m + 1 \\ &= 2(m^2 + m) + 1 \end{aligned}$$

Since 2 times any integer (even or odd) yields an even integer, it follows that $2(m^2 + m)$ is an even integer, and hence $2(m^2 + m) + 1$ is an odd integer. Hence, the statement in answer choice (C) is false.

(D) $$\begin{aligned} n^2 - m^2 = (m + 1)^2 - m^2 &= m^2 + 2m + 1 - m^2 \\ &= 2m + 1 \end{aligned}$$

Again, since 2 times any integer (even or odd) yields an even integer, it follows that $2m$ is an even integer and $2m + 1$ is always an odd integer. Hence, the statement in answer choice (D) is correct.

43. **(C)** Since the graph represents a linear function, the slope must be constant between any two points. Based on the two given points, the slope is $\frac{11 - 6}{1 - (-2)} = \frac{5}{3}$. Using the point (4, 16) in conjunction with either $(-2, 6)$ or (1, 11), the slope would also be $\frac{5}{3}$. As an example, using (4, 16) and $(-2, 6)$, the slope is $\frac{6 - 16}{-2 - 4} = \frac{-10}{-6} = \frac{5}{3}$. If any of the points in answer choices (A), (B), or (D) were paired with either $(-2, 6)$ or with (1, 11), the slope would not be $\frac{5}{3}$.

44. **(C)** The statement "Some women enjoy shopping" is equivalent to the statement "At least one woman enjoys shopping." The negation of this statement would mean that it is not true that at least one woman enjoys shopping, which becomes the statement "No women enjoy shopping."

45. **(C)** If two lines are perpendicular, their slopes must be negative reciprocals of each other. For the equation $3x + 7y = 10$, it can be rewritten as $y = -\frac{3}{7}x + \frac{10}{7}$, so its slope is $-\frac{3}{7}$. For the equation $14x - 6y = 15$, it can be rewritten as $y = \frac{14}{6}x - \frac{15}{6} = \frac{7}{3}x - \frac{5}{2}$, so its slope is $\frac{7}{3}$. Since $\frac{7}{3}$ is the negative reciprocal of $-\frac{3}{7}$, these two lines are perpendicular. In answer choice (A), the lines are parallel, since their slopes are each $-\frac{4}{7}$. For answer choice (B), the slopes are $-\frac{5}{11}$ and $-\frac{11}{5}$, so the lines are neither parallel nor perpendicular. For answer choice (D), the slopes are $-\frac{9}{2}$ and $\frac{9}{2}$, so the lines are neither parallel nor perpendicular.

46. The correct answer is 37. First find the least common multiple of 9 and 12. Rewrite each number in prime factorization form. $9 = 3^2$ and $12 = 2^2 \bullet 3$. The least common multiple is $2^2 \bullet 3^2 = 36$. This means that 36, when divided by 9 or by 12, leaves no remainder. It follows that when 37 is divided by 9 or by 12, the remainder is 1.

47. **(B)** In account on September 1: $1{,}000\left(1+\frac{0.04}{12}\right)^{12\left(\frac{1}{12}\right)} = 1{,}003.33$

In account on October 1: $2{,}003.33\left(1+\frac{0.04}{12}\right)^{12\left(\frac{1}{12}\right)} = 2{,}010.01$

In account on November 1: $3{,}010.01\left(1+\frac{0.04}{12}\right)^{12\left(\frac{1}{12}\right)} = 3{,}020.04$

They earned \$20.04 interest.

48. The correct answer is 3, namely $J \cap \varnothing$, $J \cap J'$, and $\varnothing - J$. Note that $J \cup \varnothing$ is equivalent to *J*, $J - J'$ is equivalent to *J*, and $J - \varnothing$ is equivalent to *J*.

49. **(B)** The mean is $\frac{2 + 4 + 6 + 8 + 10 + 12}{6} = \frac{42}{6} = 7$ and the median is $\frac{6 + 8}{2} = 7$. Furthermore, there is no mode, since each number appears only once. Answer choice (A) is wrong because even though both the mean and median are 4, the number 4 is also the mode. Answer choice (C) is

wrong because the mean is $\frac{46}{6}$, whereas the median is 8. Answer choice (D) is wrong because even though both the mean and median are 5, both the numbers 5 and 7 are modes.

50. **(D)** Line: $y = mx + b$ where $m = \frac{32-4}{4-0} = 7$ and $b = 4$ so $y = 7x + 4$.

Line evaluated at $x = -3$: $y = 7(-3) + 4 = -17$.

Exponential evaluated at $x = -3$: $y = 2^{-3+1} = 2^{-2} = \frac{1}{2^2} = \frac{1}{4}$.

Difference: $\frac{1}{4} - (-17) = \frac{1}{4} + 17 = \frac{69}{4}$.

51. **(A)** There are 25 − 10 = 15 green books. The number of green math books is 12 − 4 = 8. Then the number of green non-math books is 15 − 8 = 7.

52. **(B)** Set up the inequality in mathematical symbols and solve.

$$3x - 4 > 6$$
$$3x > 6 + 4$$
$$3x > 10$$
$$x > \frac{10}{3}$$

53. **(C)** 0.64 − 0.14 = 0.5 so a half pound is one standard deviation below the mean. Since 68% of the data in a normal distribution lies within one standard deviation of the mean, 34% of the data lies between the mean and one standard deviation below the mean. So 34% + 50% = 84% of the apples are above a half pound. 0.84(5,000) = 4,200 apples weigh more than a half pound.

54. **(D)** Value of stock: $9{,}000(1.06)^5 = \$12{,}044.03$

Profit before tax: $12{,}044.03 - 9{,}000 = 3{,}044.03$

Capital gains tax $= 0.15(3{,}044.03) = \$456.60$.

Value of stock after tax: $12{,}044.03 - 456.60 = \$11{,}587.43$

Profit after tax: $11{,}587.43 - 9{,}000 = \$2{,}587.43$

55. **(D)** $A = Pe^{rt} = 20{,}250(2.718)^{0.065(2.75)} = \$24{,}212.92$.

$I = 24{,}213 - 20{,}250 = \$3{,}963$

56. **(C)** When the radius is doubled, the area is multiplied by 4. To check this, suppose the radius were 3. The area is $(\pi)(3^2) = 9\pi$. If the radius doubles to 6, the area becomes $(\pi)(6^2) = 36\pi$, which is four times as large.

57. **(D)** The probability that it will not rain today is $1 - .8 = .2$, and the probability that Ken will not go bowling today is $1 - .6 = .4$. Since these events are independent, the probability that both will occur is given by $(.2)(.4) = .08$.

58. **(B)** For $f(x) = \frac{2}{x}$, the domain is only restricted by the fact that x cannot equal zero. The value of $\frac{2}{x}$ is any number except zero, since the quotient of 2 and a non-zero number must be a non-zero number. Answer choice (A) is wrong because the domain and range are non-negative numbers. Answer choice (C) is wrong because the range is non-negative numbers, but the domain is all real numbers. Answer choice (D) is wrong because the domain and range are all real numbers.

59. The correct answer is 11.5. The perimeter of the equilateral triangle is $(13)(3) = 39$, which is also the perimeter of the rectangle. The formula for the perimeter of a rectangle is $P = 2L + 2W$, where L is the length and W is the width. We know that the width is 8, so substitution into the perimeter formula gives $39 = 2L + (2)(8)$. This leads to $2L = 39 - 16 = 23$. Thus, $L = 11.5$.

60. **(D)** The new point is at the upper left. It would act as a magnet and the slope would become less steep. Since this point is far from the others, the association would become weaker in strength.

ANSWER SHEETS

Practice Test 1
Practice Test 2

PRACTICE TEST 1

Answer Sheet

1. A B C D
2. A B C D
3. A B C D
4. [ ]
5. A B C D
6. A B C D
7. A B C D
8. A B C D
9. A B C D
10. A B C D
11. A B C D
12. A B C D
13. A B C D
14. A B C D
15. A B C D
16. A B C D
17. A B C D
18. [ ]
19. A B C D
20. A B C D
21. A B C D
22. A B C D
23. A B C D
24. A B C D
25. A B C D
26. A B C D
27. A B C D
28. A B C D
29. A B C D
30. [ ]
31. A B C D
32. A B C D
33. A B C D
34. A B C D
35. A B C D
36. A B C D
37. A B C D
38. A B C D
39. A B C D
40. A B C D
41. [ ]
42. A B C D
43. A B C D
44. A B C D
45. A B C D
46. A B C D
47. A B C D
48. A B C D
49. A B C D
50. A B C D
51. A B C D
52. [ ]
53. A B C D
54. A B C D
55. A B C D
56. A B C D
57. [ ]
58. A B C D
59. [ ]
60. A B C D

PRACTICE TEST 2

Answer Sheet

1. Ⓐ Ⓑ Ⓒ Ⓓ
2. Ⓐ Ⓑ Ⓒ Ⓓ
3. Ⓐ Ⓑ Ⓒ Ⓓ
4. Ⓐ Ⓑ Ⓒ Ⓓ
5. Ⓐ Ⓑ Ⓒ Ⓓ
6. Ⓐ Ⓑ Ⓒ Ⓓ
7. Ⓐ Ⓑ Ⓒ Ⓓ
8. Ⓐ Ⓑ Ⓒ Ⓓ
9. Ⓐ Ⓑ Ⓒ Ⓓ
10.
11. Ⓐ Ⓑ Ⓒ Ⓓ
12. Ⓐ Ⓑ Ⓒ Ⓓ
13.
14. Ⓐ Ⓑ Ⓒ Ⓓ
15. Ⓐ Ⓑ Ⓒ Ⓓ
16. Ⓐ Ⓑ Ⓒ Ⓓ
17. Ⓐ Ⓑ Ⓒ Ⓓ
18. Ⓐ Ⓑ Ⓒ Ⓓ
19. Ⓐ Ⓑ Ⓒ Ⓓ
20. Ⓐ Ⓑ Ⓒ Ⓓ
21. Ⓐ Ⓑ Ⓒ Ⓓ
22. Ⓐ Ⓑ Ⓒ Ⓓ
23. Ⓐ Ⓑ Ⓒ Ⓓ
24. Ⓐ Ⓑ Ⓒ Ⓓ
25. Ⓐ Ⓑ Ⓒ Ⓓ
26. Ⓐ Ⓑ Ⓒ Ⓓ
27. Ⓐ Ⓑ Ⓒ Ⓓ
28. Ⓐ Ⓑ Ⓒ Ⓓ
29. Ⓐ Ⓑ Ⓒ Ⓓ
30. Ⓐ Ⓑ Ⓒ Ⓓ
31. Ⓐ Ⓑ Ⓒ Ⓓ
32. Ⓐ Ⓑ Ⓒ Ⓓ
33. Ⓐ Ⓑ Ⓒ Ⓓ
34. Ⓐ Ⓑ Ⓒ Ⓓ
35. Ⓐ Ⓑ Ⓒ Ⓓ
36. Ⓐ Ⓑ Ⓒ Ⓓ
37. Ⓐ Ⓑ Ⓒ Ⓓ
38.
39. Ⓐ Ⓑ Ⓒ Ⓓ
40. Ⓐ Ⓑ Ⓒ Ⓓ
41. Ⓐ Ⓑ Ⓒ Ⓓ
42. Ⓐ Ⓑ Ⓒ Ⓓ
43. Ⓐ Ⓑ Ⓒ Ⓓ
44. Ⓐ Ⓑ Ⓒ Ⓓ
45. Ⓐ Ⓑ Ⓒ Ⓓ
46.
47. Ⓐ Ⓑ Ⓒ Ⓓ
48.
49. Ⓐ Ⓑ Ⓒ Ⓓ
50. Ⓐ Ⓑ Ⓒ Ⓓ
51. Ⓐ Ⓑ Ⓒ Ⓓ
52. Ⓐ Ⓑ Ⓒ Ⓓ
53. Ⓐ Ⓑ Ⓒ Ⓓ
54. Ⓐ Ⓑ Ⓒ Ⓓ
55. Ⓐ Ⓑ Ⓒ Ⓓ
56. Ⓐ Ⓑ Ⓒ Ⓓ
57. Ⓐ Ⓑ Ⓒ Ⓓ
58. Ⓐ Ⓑ Ⓒ Ⓓ
59.
60. Ⓐ Ⓑ Ⓒ Ⓓ

Glossary

Absolute inequality: An absolute inequality for the set of real numbers means that for any real value for the variable, x, the sentence is always true.

Absolute value: The absolute value of a number is the distance the number is from the zero point on the number line. The absolute value of a number or an expression is always greater than or equal to zero (i.e., nonnegative).

Acute triangle: A triangle whose interior angles are all acute.

Adding: Increasing in amount, number, or degree.

Addition: A mathematical process to combine numbers and/or variables into an equivalent quantity, number, or algebraic expression.

Addition property of inequality: For all numbers a, b, and c, the following are true: (1) If $a > b$, then $a + c > b + c$ and $a - c > b - c$; (2) If $a < b$, then $a + c < b + c$ and $a - c < b - c$. In other words, if the same number or expression is added or subtracted from both sides of a true inequality, the new inequality is also true.

Additive inverse: The opposite of a given number.

Altitude: The height of an object or point in relation to sea level or ground level.

Altitude of the trapezoid: The distance between the bases of a trapezoid.

Angle bisector: The division of something into two equal or congruent parts, usually by a line, which is then called a bisector.

Antecedent: In the compound statement "if a, then b," statement (a) is called the antecedent.

Arc: The portion of a circle cut off by a central angle.

Area of a circle: The area of a circle is found using the formula $A = \pi r^2$.

Area of a square: The area of a square is found using the formula $A = s^2$.

Area of a triangle: The area of a triangle is found using the formula $A = \frac{1}{2}bh$.

Associative property of addition: The sum of any three real numbers is the same, regardless of the way they are grouped.

Associative property of multiplication: The product of any three real numbers is the same, regardless of the way they are grouped.

Bar graphs: A graph that uses horizontal or vertical bars to display countable data.

Base: A number used as a repeated factor.

Base angles: A pair of angles that include only one of the parallel sides.

Base of the triangle: The base of a triangle can be any side of a triangle, but specifically it is the side that is perpendicular to the height.

Bell-shaped graph: A graph that is approximately normal, i.e., mean = median = mode.

Biconditional: The statement of the form "p if and only if q."

Bimodal: When a set of data has two modes.

Bivariate data: Shows the relationship between two variables.

Cartesian product: Given two sets M and N, the Cartesian product, denoted as $M \times N$, is the set of all ordered pairs of elements in which the first component is a member of M and the second component is a member of N.

Center: The middle point of a circle or sphere, equidistant from every point on the circumference or surface.

Central angle: An angle whose vertex is at the center of a circle and whose sides are radii.

Central tendency: Mean, median, and mode, which describe the tendency (the "middle" or the "center") of a set of data.

Chord: A line segment joining two points on a circle.

Circle: The set of all points in a plane that are equidistant from a fixed point called the center.

Circle graphs: A graph that uses sections of a circle to show how portions of a set of data compare with the whole and with other parts of the data.

Circumference: The length of the outer edge of a circle.

Circumscribed circle: A circle passing through all the vertices of a polygon.

Closed interval: A set of real numbers that contains its endpoints.

Combination: An arrangement of items, events, or people from a set, without regard to the order.

Commutative property of addition: The sum of two real numbers is the same even if their positions are changed.

Commutative property of multiplication: The product of two real numbers is the same even if their positions are changed.

Complex number: A number of the form $a + bi$ where a and b are real numbers.

Composite function: Composition of functions is the process of combining two functions where one function is performed first and the result of which is substituted in place of each x in the other function.

Composite numbers: The set of integers, other than 0 and 1, that are not prime.

Concentric circles: Circles that have the same center and unequal radii.

Conclusion: The implication, "then b"; the end or finish of an event or process.

Conditional inequality: An inequality whose validity depends on the values of the variables in the sentence; that is, certain values of the variables will make the sentence true and others will make it false.

Conditional probability: The probability that event B occurs, given that event A has occurred.

Conditional statement: The compound statement "if a, then b."

Congruent circles: Circles whose radii are congruent.

Conjunction: A statement using "and."

Conjunction operator: The symbol for "and" in a conjunction ($\wedge$).

Consecutive angles: Two angles that have their vertices at the endpoints of the same side of a parallelogram.

Consecutive integers: The set of integers that differ by 1: $\{n, n + 1, n + 2, \ldots\}$ (n = an integer).

Consequent: In the compound statement "if a, then b," statement b is called the consequent.

Consistent: A system of equations is said to be consistent when it has at least one ordered pair that satisfies both equations.

Contrapositive: The contrapositive of an implication is formed by negating each statement, then swapping the order of the negated statements.

Converse: The converse of an implication is formed by switching the hypothesis and conclusion of a conditional statement.

Counting rule: If one experiment can be performed in m ways, and a second experiment can be performed in n ways, then there are $m \times n$ distinct ways both experiments can be performed in this specified order. The counting principle can be applied to more than two experiments.

Data analysis: The process of evaluating data using analytical and logical reasoning; often involves putting numerical values into picture form, such as bar graphs, line graphs, and circle graphs.

Deduction: The last part of a syllogism is a statement to the effect that the general statement which applies to the group also applies to the individual.

Deductive reasoning: The technique of employing a syllogism to arrive at a conclusion.

De Morgan's laws for sentences: Two laws, one stating that the denial of the conjunction of a class of implications is equivalent to the disjunction of the denials of an implication, and the other stating that the denial of the disjunction of a class of implications is equivalent to the conjunction of the denials of the implications.

Dependent equations: Equations that represent the same line; therefore, every point on the line of a dependent equation represents a solution.

Dependent variable: A variable (often denoted by y) whose value depends on that of another.

Diagonal: A line segment joining any two nonconsecutive sides of a polygon.

Diameter: A chord that passes through the center of the circle.

Difference of two sets: The difference of two sets, A and B, written as $A - B$, is the set of all elements that belong to A but do not belong to B.

Direction: The direction of a scatter plot tells what happens to the response variables as the explanatory variable increases.

Discriminant: The value of $b^2 - 4ac$ in a quadratic equation.

Disjoint: Two sets A and B are disjoint if they have no elements in common (their intersection is the null set).

Disjunction: Two statements a and b, shown by the compound statement "a or b."

Disjunction operator: The symbol for "or" in a conjunction ($\vee$).

Distributive laws: The property that states that multiplying a sum by a number is the same as multiplying each addend by the number and then adding the products. The distributive property says that if a, b, and c are real numbers, then: $a \times (b + c) = (a \times b) + (a \times c)$.

Distributive property: A theorem asserting that one operator can validly be distributed over another.

Domain: The set of all the values of x in a relation.

Element: Each individual item belonging to a set is called an element or member of that set.

Empty set: A set with no members.

Equal sets: Two sets are equal if they have exactly the same elements.

Equation: Two algebraic expressions separated by an equal sign that means that the two sides have equal value.

Equiangular: A triangle with each angle equaling 60 degrees.

Equilateral triangle: A triangle in which all three sides are equal.

Equivalent expressions: Expressions such as $x + 7$ and $7 + x$, which have equal values for any value of x.

Equivalent inequalities: Inequalities that have the same solution set.

Equivalent sets: Two sets are equivalent if they have the same number of elements.

Even integers: The set of integers divisible by 2: $\{\ldots, -4, -2, 0, 2, 4, 6, \ldots\}$.

Explanatory variable: Used to predict the response variable.

Exponent: Power that indicates the number of times the base appears when multiplied by itself.

Exponential function: Functions that contain a variable in an exponent.

Exterior angle: An angle formed outside a triangle by one side of the triangle and the extension of an adjacent side.

Factor: Any counting number that divides into another number with no remainder is called a factor of that number.

Factor theorem: If $x = c$ is a solution of the equation $f(x) = 0$, then $(x - c)$ is a factor of $f(x)$.

Factorial: The factorial of an integer is the product of that number and all the integers less than it down to 1, or n!

Finite set: A set where the number of its elements can be counted.

Fraction: Rational number which consist of a numerator (on the top) and a denominator (on the bottom).

Function: A relation that assigns each input exactly one output; any process that assigns a single value of y to each number of x.

Fundamental counting principle: If one experiment can be performed in m ways, and a second experiment can be performed in n ways, then there are $m \times n$ distinct ways both experiments can be performed in this specified order.

Fundamental theorem of algebra: Every polynomial equation $f(x) = 0$ of degree greater than zero has at least one root either real or complex.

Graph of an inequality: One variable is represented by either a ray or a line segment on the real number line.

Half-open interval: A continuous set of real numbers that contains only one endpoint.

Height: A line segment from a vertex of the triangle perpendicular to the opposite side.

Histogram: A graph that displays the frequency of a set of data in equal-size intervals, or ranges of data; is an appropriate display for quantitative data; is used primarily for continuous data, but it may be used for discrete data that have a wide spread.

Hypotenuse: The side opposite the right angle in a right triangle.

Hypothesis: In the compound statement "if *a*, then *b*," "if *a*" is called the hypothesis.

Identity property of addition: The sum of zero and any number is that number.

Identity property of multiplication: The product of any number and 1 is that number.

Imaginary number: The square root of -1.

Implication: The compound statement "if *a*, then *b*," also called a conditional statement.

Improper fractions: A fraction that has a numerator greater than or equal to its denominator.

Inconsistent: A system of equations with no solution.

Independent: The outcome of one event does not influence the outcome of the other event.

Independent variable: A variable whose value does not depend on the value of another variable.

Inequality: A mathematical statement that two quantities are not or may not be equal.

Infinite set: Any set that is not finite, that is, has a countless number of elements.

Inscribed angle: An angle whose vertex is on the circle and whose sides are chords of the circle.

Integers: The set of numbers represented as $\{..., -3, -2, -1, 0, 1, 2, 3, ...\}$.

Interior angle: An angle formed by two sides of a triangle and includes the third side within its collection of points.

Intersection of two sets: The intersection of two sets A and B, denoted $A \cap B$, is the set of all elements that belong to both A and B.

Intuition: Intuition is the process of making generalizations on insight.

Inverse: A function that has the range of the original function as its domain and the domain of the original function as its range.

Irrational numbers: A number that cannot be expressed as the ratio of two integers.

Isosceles trapezoid: A trapezoid whose nonparallel sides are equal.

Isosceles triangle: A triangle that has at least two equal sides.

Legs: The two sides that are not the hypotenuse of a right triangle.

Line graphs: Data display that shows points connected by line segments; often used to display data that change over time.

Line of centers: The line passing through the centers of two (or more) circles.

Linear equation: An equation with a graph that is a straight line.

Logarithmic function: Equations written in the form $y = \log_b x$ where $b > 0$ and $b \neq 1$.

Logical equivalence: Statements that are either both true or both false.

Logical value: A sentence X is true (or T) if X is true, and false (or F) if X is false.

Logically equivalent: Two sentences are logically equivalent if and only if it is impossible for one of the sentences to be true while the other sentence is false; that is, if and only if it is impossible for the two sentences to have different truth values.

Logically false: A sentence is logically false if and only if it is impossible for it to be true; that is, the sentence is inconsistent.

Logically indeterminate: A sentence is logically indeterminate (contingent) if and only if it is neither logically true nor logically false.

Logically true: A sentence is logically true if and only if it is impossible for it to be false; that is, the denial of the sentence is inconsistent.

Major premise: The first part is a general statement concerning a whole group.

Mean: The sum of the data in a data set, divided by the number of items in the set.

Median: The middle number of a data set when the data are in numerical order.

Median of a trapezoid: The line joining the midpoints of the nonparallel sides.

Member: Each individual item belonging to a set.

Midline: The line segment that joins the midpoints of two sides of a triangle; a median line of a trapezoid.

Minor premise: The second part is a specific statement which indicates that a certain individual is a member of that group.

Mode: The number or element in a data set that appears most often in the set.

Multiple: Any number that can be divided by another number with no remainder.

Multiplicative inverse: The reciprocal of a number.

Multiplicity: Number of times that $(x - r)$ is a factor of the equation.

Natural numbers: The set of counting numbers expressed as $\{1, 2, 3, ...\}$. This set is identical to the set of whole numbers, less the number zero. Natural numbers are not negative.

Negation: Given statement q, the negation is "not q."

Negation operation: If X is a sentence, then $\sim X$ represents the negation, the opposite, or the contradiction of X; $\sim$ is called the negation operation on sentences.

Negative integers: The set of integers starting with –1 and decreasing: {–1, –2, –3, …}.

Null set: A set with no members.

Number line: A line with equally spaced tick marks named by numbers.

Obtuse triangle: A triangle that has one obtuse angle (greater than 90 degrees).

Odd integers: The set of integers not divisible by 2: {… , –3, –1, 1, 3, 5, 7, …}.

Open interval: A continuous set of real numbers that does not contain its endpoints.

Order property of real numbers: If x and y are real numbers, then one and only one of the following statements is true: $x > y$, $x = y$, or $x < y$.

Parallel: Given two linear equations in x, y, their graphs are parallel lines if their slopes are equal.

Parallelogram: A quadrilateral whose opposite sides are parallel.

Permutation: An arrangement of a group, or set, of objects in a particular order.

Perpendicular: If the slopes of the graphs of two lines are negative reciprocals of each other, the lines are perpendicular to each other.

Perpendicular bisector: A line that bisects and is perpendicular to a side of a triangle.

Plane geometry: Refers to two-dimensional shapes (that is, shapes that can be drawn on a sheet of paper), such as triangles, parallelograms, trapezoids, and circles.

Point of tangency: A line that has one and only one point of intersection with a circle is called a tangent to that circle, and their common point.

Polygon: Any closed figure with straight line segments as sides.

Power: Indicates the number of times the base appears when multiplied by itself.

Premise: The compound statement "if a, then b" is called a conditional statement or an implication. "If a" is called the hypothesis or premise of the implication.

Prime numbers: The set of positive integers greater than 1 that are divisible only by 1 and themselves: {2, 3, 5, 7, 11, …}.

Product: The answer when two or more numbers are multiplied together.

Proper fractions: Numbers between –1 and 1; the numerator is less than the denominator.

Pythagorean Theorem: The lengths of the three sides of a right triangle are related by the formula $a^2 + b^2 = c^2$.

Quadratic equation: An equation in which the highest power of an unknown quantity is two.

Quadratic formula: If the quadratic equation does not have obvious factors, the roots of the equation can always be determined by the quadratic formula in terms of the coefficients a, b, and c:

$$x = \frac{-b \pm \sqrt{b^2 - 4ac}}{2a}.$$

Quadrilateral: Any polygon with four sides.

Radius: Fixed distance from the center of the circle to any point on the circle.

Range: The set of all the values of y in a relation; the difference between the greatest number in a list of numbers and the least.

Rational numbers: The set of all terminating and repeating decimals; is of the form, where p and q are integers, and q is not equal to zero.

Real number: Any positive or negative number; includes all integers and all rational and irrational numbers.

Real part: The real and imaginary parts of a complex number z are respectively the first and the second elements of the ordered pair.

Reciprocal: One of two numbers whose product is 1; two numbers are reciprocals of each other if their product equals 1.

Rectangle: A parallelogram with right angles.

Reflection: Movement of a figure to a new position by flipping it over a line.

Remainder theorem: If a is any constant and if the polynomial $p(x)$ is divided by $(x - a)$, the remainder is $p(a)$.

Response variable: Measures the outcomes that have been observed.

Rhombus: A parallelogram that has two adjacent sides that are equal.

Right triangle: A triangle that has a right angle.

Rotation: A function moves each point P to a new point P' so that $OP = OP'$.

Scalene triangle: A triangle that has no equal sides.

Scatter plots: A graph that shows data points on a coordinate grid.

Secant: A line that intersects a circle in two points.

Sense: Two inequalities are said to have the same sense if their signs of inequality point in the same direction.

Set: A collection of items.

Shape of a curve: A plot is usually classified as linear or nonlinear (curved).

Side of a triangle: A line segment whose endpoints are the vertices of two angles of the triangle.

Similar: A one-to-one correspondence between the vertices of a polygon such that all pairs of corresponding angles are congruent and the ratios of the measures of all pairs of corresponding sides are equal.

Simultaneous equations: Equations involving two or more unknowns that are to have the same values in each equation.

Skewed: The graph will either be skewed to the left, skewed to the right, or approximately normal. A skewed distribution of data has one of its tails longer than the other.

Slope-intercept form: The equations will be in the form $y = mx + b$, where m is the slope and b is the intercept on the y-axis.

Solution: A number that makes the equation true when it is substituted for the variable.

Square: A rhombus with a right angle.

Standard deviation: The standard deviation is a numerical value used to indicate how widely individuals in a group vary; the square root of variance.

Statement: A sentence that is either true or false, but not both.

Stemplot: Also called stem-and-leaf plot, can be used to display univariate data.

Strength of correlation: How tight or spread out the points of a scatter plot are.

Subset: Given two sets A and B, A is said to be a subset of B if every member of set A is also a member of set B.

Substitution: A method of solving systems of equations used to find an expression for the value of a variable from one equation and then substitute the expression for that variable in the other equation.

Subtraction: Process or skill of taking one number or amount away from another.

Syllogism: An arrangement of statements that would allow you to deduce the third one from the preceding two.

Tangent: A line that has one and only one point of intersection with a circle.

Transitive property of inequalities: If a, b, and c are real numbers with $a > b$, and $b > c$, then $a > c$. Also, if $a < b$ and $b < c$, then $a < c$.

Translation: A function will move each point of the function a specific number of units left or right, then up or down.

Trapezoid: For quadrilateral with two and only two parallel sides.

Triangle: A closed three-sided geometric figure.

Truth table: For sentence X, the exhaustive list of possible logical values of X.

Union of two sets: The union of two sets A and B, denoted $A \cup B$, is the set of all elements that are either in A or B or both.

Universal set: A set from which other sets draw their members.

Valid: The truth of the premises means that the conclusions must also be true.

Variability: The quality of being subject to variation.

Variance: The variance tells us how much variability exists in a distribution. It is the "average" of the squared differences between the data values and the mean.

Venn diagram: A visual way to show the relationships among or between sets that share something in common.

Vertex: For a parabola the point where it intersects the axis of symmetry; the highest or lowest point on a parabola.

Vertices: For a hyperbola the line through the foci that intersects the graph at two points; the points where the lines of the equations of two constraints intersect.

Whole numbers: The set of whole numbers can be expressed as $\{0, 1, 2, 3, ...\}$. This is the set of natural numbers and zero. Whole numbers are not negative.

Index

Notes

Notes

Notes

Notes

Notes

Notes